Dire Strait? Military Aspects of the China-Taiwan Confrontation and Options for U.S. Policy

David A. Shlapak
David T. Orletsky
Barry A. Wilson

*Supported by the
Smith Richardson Foundation*

National Security Research Division

RAND

The research described in this report was sponsored by the Smith Richardson Foundation. The research was conducted within the International Security and Defense Policy Center of RAND's National Security Research Division.

Library of Congress Cataloging-in-Publication Data

Shlapak, David A.
    Dire strait? ; military aspects of the China–Taiwan confrontation and options for U.S. policy / David A. Shlapak, David T. Orletsky, Barry Wilson.
      p. cm.
    "MR-1217-SRF."
    Includes bibliographical references.
    ISBN 0-8330-2897-9
    1. Taiwan—Military policy. 2. China—Military policy. 3. United States—Military policy. I. Orletsky, David T., 1963– II. Wilson, Barry, 1959– III. Title.

UA853.T28 S55 2000
355'.03355124'9—dc21

                                                                                                                    00-062657

RAND is a nonprofit institution that helps improve policy and decisionmaking through research and analysis. RAND® is a registered trademark. RAND's publications do not necessarily reflect the opinions or policies of its research sponsors.

© Copyright 2000 RAND

All rights reserved. No part of this book may be reproduced in any form by any electronic or mechanical means (including photocopying, recording, or information storage and retrieval) without permission in writing from RAND.

Published 2000 by RAND
1700 Main Street, P.O. Box 2138, Santa Monica, CA 90407-2138
1200 South Hayes Street, Arlington, VA 22202-5050
RAND URL: http://www.rand.org/
To order RAND documents or to obtain additional information, contact Distribution Services: Telephone: (310) 451-7002;
Fax: (310) 451-6915; Internet: order@rand.org

PREFACE

Even a half century after the birth of the People's Republic of China (PRC), the Taiwan Strait remains the locus of one of the most dangerous military confrontations in the world. In recent years, a series of Chinese military exercises coupled with the ongoing modernization of the People's Liberation Army (PLA) have seemed to raise the stakes in this long-standing staredown and likewise increased its visibility, especially in the United States.

Until 1979, the United States was Taiwan's primary security partner. Today, it remains linked to the island by both force of law and a natural affinity toward a rapidly democratizing polity embedded in a vibrant market economy. But Washington at the same time is pursuing improved relations with Beijing as well as encouraging the PRC's deeper integration with the international system at large. Because the status of Taiwan may be China's single most neuralgic point, the United States is compelled to perform a delicate balancing act— attempting to fulfill its obligations and inclinations toward ensuring the Republic of China's (ROC) survival without making an enemy of the mainland.

This report looks at the near-term military balance between China and Taiwan. Mixing quantitative and qualitative analysis, it explores a range of key factors that affect the ROC's self-defense capabilities and suggests ways that the United States can effectively contribute to improving the odds in Taipei's favor.

This report was written as part of a project on assessing Taiwanese defense needs, sponsored by the Smith Richardson Foundation. Research for the report was conducted within the International

Security and Defense Policy Center of RAND's National Security Research Division (NSRD), which conducts research for the U.S. Department of Defense, for other U.S. government agencies, and for other institutions. Publication of this report was supported in part by the Strategy and Doctrine program of Project AIR FORCE.

CONTENTS

Preface	iii
Figures	vii
Tables	ix
Summary	xi
Acknowledgments	xxi
Abbreviations	xxiii

Chapter One
    INTRODUCTION 1
    Confrontation in the Taiwan Strait 1
    The U.S. Role 2
    The Purpose of This Study 3
    Structure of This Report 4

Chapter Two
    SCENARIO AND APPROACH 7
    A Chinese Invasion of Taiwan 7
    Analytic Structure and Methodology 9
        Scoping the Problem 9
        Air War Methodology 12
        Naval War Methodology 18
        Caveats 18
    Orders of Battle 19
        Air, Air Defense, and Missile Forces 19
        Naval Forces 20
        Command and Control 23

Playing Out the Scenario	24
Overview	24
The War in the Air	24
The War at Sea	30

Chapter Three

ISSUES AND IMPLICATIONS	31
Air Superiority	31
Base Operability	31
Advanced Air Weapons	34
Training Quality	36
The Value of U.S. Involvement	38
Maritime Superiority	40
The ASW Dilemma	40
Maintaining a Credible Antisurface Warfare Capability	42
The U.S. Role	43
Summing Up	45

Chapter Four

RECOMMENDATIONS AND CONCLUDING REMARKS	47
U.S. Support Is Vital to Taiwan's Security	47
Small Increments of U.S. Assistance Could Turn the Tide	48
Supporting Taiwanese Modernization: The Israel Model	49
Air Defense C^2	51
Information and Intelligence Sharing	51
Interoperability: The Critical Link	52
China as a Sanctuary?	54
Looking Beyond 2005	55
Final Thoughts	56

Appendix

A. Some Thoughts on the PRC Missile Threat to Taiwan	59
B. Overview of the JICM	63
References	85

FIGURES

2.1	Overall Outcomes	26
3.1	Effects of Reductions in ROCAF Sortie Rates	32
3.2	Effects of BVR Capabilities on Case Outcome	35
3.3	Effects of ROC Training on Case Outcome	37
3.4	Overall Effect of U.S. Involvement	39
3.5	Effect of U.S. Involvement on Air Outcomes, Advanced Threat	40
B.1	Fraction of Package with First Shot	69
B.2	D-Day Sorties	76
B.3	D-Day Sortie Losses	77
B.4	D-Day Aircraft Losses and Sorties	77

TABLES

2.1	PLAAF Forces Committed to Taiwan Contingency	14
2.2	Chinese Missile Forces	15
2.3	Cases for Exploratory Analysis	18
2.4	ROCAF Composition	19
2.5	Taiwan Surface-to-Air Order of Battle	20
2.6	Taiwanese Naval Order of Battle	21
2.7	Chinese Naval Order of Battle	21
2.8	Impact of Parameters on "Red" Outcomes	27
B.1	Mission Packages	74
B.2	Package Timing	75
B.3	Aircraft Data	79
B.4	Engagement Rates and First Shots	80
B.5	Air-to-Air Weapon Data	80
B.6	Training Factors	81
B.7	Air-to-Air Weapon Loads	81
B.8	Air-to-Air Exchange Rates	82
B.9	Air-to-Ground Weapon Data	83
B.10	Air-to-Ground Weapon Loads	83
B.11	Ground-to-Air Weapon Data	83

SUMMARY

OVERVIEW

As the new century dawns, the Taiwan Strait is the locus of one of the world's most dangerous flashpoints. Two entities share the name of "China": one, the most populous country in the world, is a gargantuan and unique hybrid of Communist ideology and capitalist appetite, while the other is a tiny island republic of great wealth and uncertain international status. And across the narrow barrier of the Taiwan Strait, these two powers—the People's Republic of China (PRC) and the Republic of China (ROC)—stare at each other.

The United States plays an interesting role in this pas de deux, part observer and part participant. For 30 years after 1949, it was Taiwan's principal patron, maintaining a mutual defense treaty with the ROC. When the 1970s brought a "normalization" of relations between Washington and Beijing, this era of close cooperation ended. Since 1979, the U.S. government has maintained a calculated ambiguity in its policy toward the deadlock over Taiwan's status. This balancing act has been complicated recently by such events as China's 1995 and 1996 missile tests, in the wake of which Taiwan's security situation has gained new visibility in Washington, where concerns have been raised about whether the United States is doing enough to ensure the island's self-defense capabilities.

This monograph reports the results of a project that examined the military dimensions of the confrontation between China and Taiwan.[1] Using a mixture of qualitative and quantitative analysis, we have done two things:

- Identified a handful of issues that appear crucial in helping Taiwan maintain an adequate defensive posture vis à vis the PRC, and
- Developed a set of recommendations for steps the United States might take to assist Taipei in dealing with those issues.

SCENARIO AND APPROACH

Although coercive scenarios (e.g., limited missile strikes) are usually regarded as the more likely form of Chinese use of force against Taiwan, we assessed the more extreme case of an outright air and amphibious invasion of the island. We chose to focus on this challenging contingency for six reasons.

- Some analysts argue—in contrast to the conventional wisdom—that "immediate and full-scale invasion" is the most likely form of conflict between the two sides.
- As the "worst-case" scenario, it is of interest to military planners whose responsibility it is to deter potential adversaries from dangerous courses of action.
- The possibility of a direct Chinese invasion of Taiwan—and expectations regarding the outcome of such an attack—is important in shaping overall perceptions of the balance between the two sides.
- The seizure and holding of the island is the only alternative that guarantees Beijing's control when hostilities end. So, in some

---

[1] That this report focuses on military issues should not be interpreted as suggesting that the crux of the China-Taiwan issue is military; neither do the authors believe that military means are the only or even the most likely way of resolving the dispute. Our given task has been to examine the balance of power across the strait, not to document, explicate, or predict the complex political dynamics at the heart of the differences between Taipei and Beijing. We recognize that a strongly deterrent Taiwanese posture is only one part—albeit a vital one—of the equation for maintaining peace and stability on the strait and in East Asia.

sense, the credibility of the invasion threat underwrites the other, lower-level options, such as limited missile strikes or maritime harassment.

- While it seems unlikely that China would undertake such a desperate gamble, it is important to think through the manner in which the People's Liberation Army (PLA) might essay the operation and what steps would be needed to defeat it. After all, it was always terribly unlikely that the Soviet Union would launch a massive nuclear attack on the United States. Still, hundreds if not thousands of war games, exercises, and analyses were invested in exploring the "what-ifs" of that contingency.

- An invasion scenario incorporates a number of elements that could be components of other coercive strategies directed against Taiwan. Perhaps most significant is the employment of conventionally armed surface-to-surface missiles (SSMs) against targets in Taiwan.

Analytically, an invasion campaign can be divided into four segments:

- In the first phase, the two sides would fight for air superiority.
- The second phase, which could begin simultaneously with the first, would be a struggle for maritime control of the strait.
- Followup air strikes would focus on "softening up" the island's defenses.
- The fourth phase would involve actual landing operations and could include amphibious landings, paratroop assaults, and heliborne attacks.

Our attention is focused mainly on the battle for air superiority and, secondarily, on the contest for control of the seas. Control of the air and control of the sea are absolute prerequisites for a successful amphibious and/or airborne assault. This may be particularly true in the context of a PRC attack on Taiwan. The People's Liberation Army Navy (PLAN) owns enough amphibious lift to move about a division of troops at a time, hardly enough to establish and sustain a firm foothold in the face of determined Taiwanese resistance. Therefore, many analyses picture a kind of "Dunkirk in reverse," with China employing numerous commercial vessels to transport troops, equipment, and supplies across the strait. Such an operation, involving unarmed merchant shipping, would be sheer folly unless China had secured almost uncontested dominance of the air and sea. Similarly, the kind of large-scale airborne and air assault operations often suggested as part of a PRC attack would be virtually suicidal unless the ROC's air defenses had been thoroughly suppressed. Finally, the surface forces of the two navies consist of warships with very limited air defense capabilities. In the absence of air superiority, the PLAN's warships would be very vulnerable to air attack in the confined waters of the strait. We therefore conclude that the battle for air superiority in particular is the linchpin of the campaign.

We chose RAND's Joint Integrated Contingency Model (JICM) as the primary modeling tool for this study; developed for the U.S. Department of Defense, JICM is a theater combat model designed to support the kind of exploratory analysis that we emphasized in this project. After preparing a database from open-source materials and making an initial set of runs to identify the factors that seemed likely to play a determining role in the outcome of the war over the strait, we conducted more than 1,700 model runs to examine the impact of seven key variables:

- The size and composition of the air forces committed to the attack by the PRC.
- Each side's possession of beyond-visual-range (BVR), "fire-and-forget" medium-range air-to-air missiles (AAMs).
- The number and quality of short- and medium-range ballistic missiles (SRBMs and MRBMs) used by the Chinese.
- The number of advanced precision-guided munitions (PGMs), such as laser-guided bombs (LGBs) and Global Positioning System (GPS)-guided weapons, in the Chinese inventory.
- The ability of the Republic of China Air Force (ROCAF) to generate combat sorties.
- The quality of the ROCAF's aircrew.
- The extent, if any, of U.S. air forces, both land and sea based, committed to Taiwan's defense.

Our more-limited analysis of the naval war was undertaken using the JICM and *Harpoon*, a computer-based simulation of maritime warfare. *Harpoon* is widely considered the best commercially available depiction of modern maritime combat. It includes representations of submarine, surface, and air warfare.

This work explores only a very limited region of what is often referred to as the "scenario space." We concentrated on one specific scenario involving one particular Chinese offensive strategy, and we selected the factors to vary based on our reading of the extant literature on the China-Taiwan balance as well as discussions with experts in the United States and elsewhere. We also focused our attention on what might be thought of as "reasonable" cases: those reflecting current capabilities, linear projections of current capabilities, and capabilities conceivably attainable within our limited time frame. As such, we present these results as *illustrative* and *indicative*, meant to highlight and illuminate certain key points that emerged from our overall analysis.

Because our notional war is set in 2005, the two sides' orders of battle consist largely of systems already present in their arsenals. We varied the size and composition of the PRC air and missile forces committed to the campaign to reflect uncertainties regarding the pace and scale of China's military modernization programs.

The analysis required many assumptions, and the problem frequently arose as to how much credit to give the protagonists for various capabilities. We decided to credit both sides with taking measures to increase their competence in critical areas. In particular, we credited the Chinese with more capability than they have actually demonstrated in conducting complex offensive operations. And we assumed that Taiwan would be able to maintain the basic functionality of its command and control (C^2) system, even under the stress of a concerted PRC attack.[2] Because of these assumptions, our analysis is less a current net assessment of actual capabilities on the two sides than it is an assessment of reasonable *potential* capabilities with given orders of battle.

---

[2] Including possible, but unmodeled, information warfare operations.

RESULTS, IMPLICATIONS, AND RECOMMENDATIONS

Our analysis suggests that any near-term Chinese attempt to invade Taiwan would likely be a very bloody affair with a significant probability of failure. Leaving aside potentially crippling shortcomings that we assumed away—such as logistics and C^2 deficiencies that could derail an operation as complex as a "triphibious" (amphibious, airborne, and air assault) attack on Taiwan—the PLA cannot be confident of its ability to win the air-to-air war, and its ships lack adequate antiair and antimissile defenses. Provided the ROC can keep its air bases operating under attack—a key proviso that we will discuss at length in the next chapter—it stands a relatively good chance of denying Beijing the air and sea superiority needed to transport a significant number of ground troops safely across the strait. Overall, the ROC achieved "good" outcomes in almost 90 percent of the cases against our best-estimate "base" PRC threat. Both in the air and at sea, attrition was extremely high on both sides.[3]

We identified seven key findings from our analysis:

- *Taiwan's air bases must remain operable so that the ROCAF's fighter force can keep up the fight against the superior numbers of the PLA Air Force (PLAAF).* We recommend increased attention to passive defense and rapid-reconstitution measures; Taiwan could learn much from NATO's response to the threat posed to its rear area by Warsaw Pact air and missile attacks in the 1970s and 1980s.

- *The ROC must maintain at least parity in advanced air-to-air weaponry.* Ideally, Taiwan would enjoy a unilateral advantage in this area. At the very least, however, the PLAAF cannot be permitted to field significant quantities of "fire-and-forget" AA-12-class weapons without Taiwan being similarly endowed with AIM-120 Advanced Medium-Range Air-to-Air Missiles (AMRAAMs). The recent decision by the U.S. government to

---

[3]For a variety of reasons, the attrition we calculated may be higher than would occur in an actual clash between China and Taiwan. Nonetheless, we believe that such a conflict would feature loss rates that would be extremely high by historical standards.

provide AMRAAMs to the ROCAF if China acquires the AA-12/R-77 is an important and welcome hedge.

- *Pilot quality may be Taiwan's ace in the hole.* PLAAF training is notoriously poor. This makes it even more important for Taiwan to ensure that their aircrews are of the highest possible caliber. Our analysis suggests that improved pilot quality may contribute more to favorable air superiority outcomes than would even sizable additions to the ROCAF's fighter force structure.

- *U.S. involvement is important now and will likely grow increasingly vital.* Even in the near term, U.S. carrier- and land-based fighters could make a combination crucial to Taiwan's defense. As the PLAAF's inventory becomes more sophisticated and capable, Taiwan's need for U.S. assistance will likewise increase.

- *Antisubmarine Warfare is a critical Taiwanese weakness.* Absent an unexpected acquisition of numerous modern attack submarines, the ROC Navy (ROCN) will have tremendous difficulty coping with China's modernizing submarine fleet. We suggest that Taiwan's navy consider keeping its main battle forces out of the strait during the initial phase of a war with the mainland.

- *Fast, stealthy missile boats and highly mobile land-based antiship missile launchers can help Taiwan exploit its inherent defensive advantages.* If adequate detection and targeting information can be provided, these weapons could prove highly lethal and relatively survivable even in the chaotic opening hours of a China-Taiwan clash.

- *Again, the U.S. role in the naval campaign could be crucial.* U.S. nuclear-powered attack submarines (SSNs) could help counter the Chinese submarine threat, U.S. surveillance capabilities could provide vital support to Taiwanese forces, and Harpoon-equipped bombers could provide early firepower key to the naval battle.

Given that it seems unlikely that Beijing will renounce its "right" to use force to compel unification, a strong Taiwanese deterrent appears to be a necessary component of continued peace on the strait. As Taiwan's most reliable friend and in keeping with the requirements of the 1979 Taiwan Relations Act, the United States will necessarily play a major role in helping the ROC maintain and enhance its defensive capabilities even as the PLA modernizes. Should deterrence fail, Taiwan may find itself in a position where its survival is dependent on some degree of direct U.S. military intervention.

Our analysis, however, suggests five key insights regarding U.S. support for Taiwan—in both peace and war—that indicate ways of enhancing deterrence across the strait. By pursuing initiatives along these lines, Taiwan's defense posture vis à vis China could be significantly enhanced with, we believe, minimal risk of destabilizing the situation.

First, *the amount of force needed to support Taiwan in the near term appears to fall considerably short of what is usually thought of in the Pentagon as that needed to prosecute a major theater war* (MTW). In our analysis, we never committed more than a single wing of land-based fighters, two carrier battle groups (CVBGs), and a dozen or so heavy bombers to the campaign—a far smaller force than the 10 fighter wing equivalents and six CVBGs that were engaged in *Desert Storm*.

In terms of arms sales and military assistance, our second recommendation is that *attention should focus on helping Taiwan get the most out of its existing inventory of advanced platforms rather than selling the ROC entirely new weapon systems.* Providing key advanced weapons, such as AMRAAM, improved sensors, and enhanced training, would be important elements of such a strategy.

Third, Taiwan's air defense C^2 network, which has been upgraded substantially in the past decade, continues to suffer from limitations in intelligence fusion and data transmission. *These shortcomings should be an important priority for rectification.* The U.S. side can encourage Taiwan to make the investments needed to ensure that the ROC's C^2 system is fully modernized and robust in the face of the kinds of threats it would likely face in a conflict with China.

Fourth, the United States is obviously and properly sensitive and selective in choosing how and when to share what kinds of information and intelligence with its friends and allies. At the same time, however, *there would appear to be enormous leverage to be gained by helping Taiwan's government and military leadership maintain an accurate picture of the strategic and tactical situation* day to day and, especially, during a crisis. A shared picture of the evolving threat would also likely make it easier for the two sides to reach agreement on arms sales and other modes of U.S.-Taiwan defense cooperation.

Finally, we wish to call attention to the *critical problem of interoperability*, should Taiwanese and U.S. forces ever find themselves required to fight side by side. This analysis assumed that the United States and Taiwan had achieved only a minimum level of interoperability, but even this may overstate the degree of cooperation that would be possible if war were to break out today. Enhancing the ease of cooperation between Taiwanese and U.S. forces—even to the extent of ensuring that the two countries' forces can merely stay out of one another's way in a crisis—is in the interests of both sides, and even small and discreet steps could be valuable.

In addition to working with Taiwan to improve the ROC's deterrent posture, the United States could begin to think through some of the operational-strategic issues that would be raised by the need to support Taiwan actively in a conflict against China. As demonstrated in Iraq and again in the Balkans, contemporary U.S. warfighting strategy typically includes large-scale strikes against command, control, and communications (C^3) facilities, air defenses, air bases, and an array of other targets in the adversary's territory. Whether or not the United States would initiate such a campaign against a nuclear-armed opponent, such as China—and, if so, what sorts of limitations would be imposed on targeting and collateral damage—is a deeply vexing question.

The need to suppress the PLA's long-range air defenses could provide the most compelling rationale for at least limited attacks on military targets in China. Neutralizing long-range "double-digit" surface-to-air missiles (SAMs) is widely regarded as a difficult tactical problem; adding the risks associated with attacking even strictly military targets within China compounds the complexity.

LOOKING BEYOND 2005

This study was exclusively focused on the near term and included only capabilities that could conceivably be fielded by 2005.

Nonetheless, our work suggests four developments on the Chinese side that appear particularly troublesome:

- Advances in information warfare capabilities that enable China to shut down Taiwan's C^2 networks more rapidly and completely.
- The deployment of hundreds or thousands of conventionally armed and highly accurate ballistic and cruise missiles that could greatly endanger the operability of Taiwan's air bases.
- Fielding of a standoff munition similar to the U.S. Joint Standoff Weapon (JSOW) that would enable the PLAAF to accurately deliver ordnance onto many Taiwanese targets from within or just outside the coverage umbrella provided by China's long-range SAMs.
- Large numbers of GPS-guided free-fall munitions (akin to the U.S. Joint Direct Attack Munitions [JDAM]) that might turn older aircraft with poorly trained pilots into reasonably effective attack platforms.

Looking toward this uncertain future, we recommend that the United States work to help Taipei improve its ability to defend key military and commercial information systems from attack. Also, with the Chinese likely to exploit GPS and Russian Global Navigation Satellite System (GLONASS) navigation satellites in the guidance modes for many future weapons, Taiwan may want to acquire the ability to jam these signals effectively over both its own territory and the strait.

ACKNOWLEDGMENTS

This work benefited from the contributions of many individuals both outside of RAND and within.

First, we wish to thank Dr. Marin Strmecki and the Smith Richardson Foundation for supporting this project. Absent their interest, this volume quite simply would not exist.

Within RAND, we want to acknowledge Jeff Isaacson, initially the program director under whom this project was undertaken, and his successor in that position, Stuart Johnson. Roger Cliff and James Mulvenon served as project leaders and were instrumental in shaping the work. Zalmay Khalilzad provided the initial impetus for the study and remained constructively interested throughout the process. Project AIR FORCE's Strategy and Doctrine Program, of which Zal is director, also provided additional funding to support the final publication of this report, for which generosity we are very grateful.

Michael Swaine and Paul Davis reviewed the draft version of this document and recommended many changes, all of which contributed to greatly improving the ultimate product; we thank them for their care, their dedication, and for not wringing our necks when we wrangled over fine and not-so-fine points of interpretation. Daniel Sheehan edited the paper, helping bring coherence to our prose. Lisa Rogers provided able administrative and editorial assistance throughout the project.

We warmly thank Rachel Swanger for her critical role in shepherding us through the final phases of the project and into the publications process. Her help was invaluable.

Finally, our thanks go to the many members of the Republic of China armed forces who assisted us both here and on our visit to Taiwan. While they must remain nameless here, we know who they are and are grateful for their help.

It is said that success has many parents while failure is an orphan. As these acknowledgments suggest, the former certainly holds true for our work here. In this case, however, failure, too, has its progenitors. As hard as we might try to pass the buck, whatever shortcomings and errors of omission or commission reside within these pages are the responsibility of the authors alone.

ABBREVIATIONS

AAM	Air-to-air missile
AAW	Antiair warfare
AI	Air interdiction
AMRAAM	Advanced Medium-Range Air-to-Air Missile
ASM	Antiship missile
ASW	Antisubmarine warfare
ATO	Air Tasking Order
AWACS	Airborne Warning and Control System
BMD	Ballistic missile defense
BVR	Beyond visual range
C^2	Command and control
CAP	Combat air patrol
CVBG	Carrier battle group
CVW	Carrier air wing
EK	Expected number of kills
ELINT	Electronic intelligence
EW	Early warning
GLONASS	Global Navigation Satellite System
GPS	Global Positioning System
IADS	Integrated air defense system
IAF	Israeli Air Force

IDF	Indigenous Defense Fighter
IFF	Identification, friend or foe
JDAM	Joint Direct Attack Munition
JICM	Joint Integrated Contingency Model
JSOW	Joint Standoff Weapon
LACM	Land-attack cruise missile
MND	Ministry of National Defense (Taiwan)
MRBM	Medium-range ballistic missile
MTW	Major theater war
NATO	North Atlantic Treaty Organization
PLA	People's Liberation Army
PLAAF	PLA Air Force
PLAN	PLA Navy
PRC	People's Republic of China
ROC	Republic of China
ROCAF	ROC Air Force
ROCN	ROC Navy
SAM	Surface-to-air missile
SARH	Semiactive radar homing
SEAD	Suppression of enemy air defenses
SIGINT	Signals intelligence
SOF	Special Operations Forces
SRBM	Short-range ballistic missile
SS	Attack submarine
SSM	Surface-to-surface missile
SSN	Nuclear-powered attack submarine
TBM	Tactical ballistic missile
TMD	Theater missile defense
TRA	Taiwan Relations Act
WMD	Weapons of mass destruction

Chapter One
INTRODUCTION

CONFRONTATION IN THE TAIWAN STRAIT

As bodies of water go, the Taiwan Strait is not the most impressive. Barely 100 miles wide at its narrowest point, no oil or mineral wealth lies below it, no fairy-tale castles of coral attract tourists to swim among a flashing rainbow of colorful tropical fish. This is a workaday stretch of the Pacific, where fishermen reel out their nets and commercial shipping goes on its prosaic way.

Yet as the new century dawns, the Taiwan Strait is the locus of one of the world's most dangerous flashpoints. Two entities share the name of "China": one, the most populous country in the world, is a gargantuan and unique hybrid of Communist ideology and capitalist appetite, while the other is a tiny island republic of great wealth and uncertain international status. And across the narrow barrier of the Taiwan Strait, these two powers—the People's Republic of China (PRC) and the Republic of China (ROC)—stare at each other.[1]

For the leadership in Beijing, Taiwan is a rebellious province whose ultimate destiny must be political and economic unification with the mainland. In Taipei, meanwhile, the ROC government neither races toward reunion nor utterly forswears it but embraces instead an uneasy status quo. Both sides manage a delicate balancing act, jug-

---

[1]This may be a good moment to dispose of an issue that might otherwise plague this discussion. For purposes of this paper, the name "China" by itself refers to the PRC; the parallel term for the ROC will be "Taiwan." This is simply a matter of terminological convenience.

gling concession and confrontation, striving to advance their respective positions in the face of their mutual distrust. Neither side seems anxious to resort to arms to resolve the question of Taiwan's status once and for all, but both are aware that such a confrontation could come to pass. Indeed, Beijing has a disquieting tendency to rattle its saber on those occasions when the Taipei government behaves in ways the Communist leadership finds offensive.

THE U.S. ROLE

The United States plays an interesting role in this pas de deux, part observer and part participant. For 30 years after 1949, it was Taiwan's principal patron, maintaining a mutual defense treaty with the ROC. When the late 1970s brought a "normalization" of relations between Washington and Beijing, this era of close cooperation ended. Since 1979, the U.S. government has maintained a calculated ambiguity in its policy toward the deadlock over Taiwan's status. While recognizing Beijing as "China" and forswearing formal diplomatic ties with the ROC, the Taiwan Relations Act enjoins Washington to "enable Taiwan to maintain a sufficient self-defense capability."[2] Each successive administration, whether Democratic or Republican, has stated repeatedly that it opposed any attempt by the mainland to effect unification by force but would support any peaceful resolution to the standoff that was mutually agreeable to the two principals.

For more than 15 years, this policy worked well. The tensions between China and Taiwan sat in the background of U.S. foreign policy concerns, overshadowed by the final struggles of the Cold War and the birth pangs of the era that emerged from the ashes of that long confrontation. In the waning years of the East-West confrontation, China was seen as a potential strategic partner against Soviet expansionism in Asia. After the USSR's collapse, China was viewed by many as either an emerging economic powerhouse or an authoritarian human-rights abuser. To the extent that Taiwan entered U.S. calculations at all, it was as a dynamic "Asian tiger" or a "newly

---

[2]U.S. Congress, 96th Congress, 1st Session, Taiwan Relations Act, Public Law 96-8, section 3(a). The full text of the TRA can be found at several locations on the Web, including: http://ait.org.tw/ait/tra.html.

industrialized country," albeit one without a recognized state to represent it on the international scene.

This relative calm was disturbed in the early 1990s when Beijing grew agitated over the rise of pro-independence political sentiments on Taiwan and the U.S. agreement to sell 150 F-16 fighters to Taipei. These simmering tensions erupted dramatically in March 1996, when, in an attempt to influence Taiwan's forthcoming presidential elections, China launched four short-range DF-15 ballistic missiles—nuclear-capable delivery vehicles—into open-ocean target areas near the island nation's two largest ports, Keelung and Kaohsiung.[3] The United States responded by deploying first one then two carrier battle groups (CVBGs) into the waters around Taiwan, though not into the strait itself. According to one commentator, "This was the largest U.S. show of force directed at China since the Straits crises of the 1950s." (Fisher, 1997, p. 178.)

While U.S. declaratory policy has not changed since these events, many observers believe that the "missile crisis" proved a turning point of sorts in Washington's perceptions of and commitment to Taiwan. The clumsiness of China's attempted coercion, contrasted with the peaceful democratic process playing out in Taiwan, undoubtedly elevated the latter's status in the eyes of many Americans and may have measurably increased the likelihood of U.S. intervention in the event of an armed clash between Beijing and Taipei. Taiwan's security situation has certainly gained new visibility in Washington, where, particularly on Capitol Hill, concerns have been raised about whether the United States is doing enough to ensure the island's self-defense capabilities.

THE PURPOSE OF THIS STUDY

Since the end of the Cold War, U.S. military planning has centered around two near-simultaneous major theater wars (MTWs), usually scenarios in the Persian Gulf and on the Korean peninsula. This focus, combined with the lack of any defense arrangements with or

---

[3]China has also conducted missile tests the previous July in apparent reaction to then-Taiwan President Lee Teng-hui's "private" visit to his U.S. alma mater, Cornell University.

pertaining to Taiwan, means that little if any formal planning has been done on how the United States might support the ROC in the event of war with the mainland. What might such a conflict look like? What are the keys to a successful defense of Taiwan? What might the United States do both before and during such a crisis to help promote Taiwan's security?[4]

This report documents a project that examined the military dimensions of the confrontation between China and Taiwan. Using a mixture of qualitative and quantitative analysis, we have done two things:

- Identified a handful of issues that appear crucial in helping Taiwan maintain an adequate defensive posture vis à vis the PRC, and
- Developed a set of recommendations for steps the United States might take to assist Taipei in dealing with those issues.

We do not claim our list is exhaustive; neither do we claim to have performed the definitive analysis of the China-Taiwan strategic equation.[5] However, we do believe that our analysis has pointed toward several very strong conclusions that should be accounted for in future U.S.-Taiwan security discussions.

STRUCTURE OF THIS REPORT

In the next chapter, we describe the scenario we used as a focusing mechanism for our analysis, our general approach, and our overall findings. Chapter Three is organized around seven issues that we assess as critical to Taiwan's near-term defense capabilities. We pre-

---

[4]The analysis in this paper is limited to conventional warfare and does not assess the possible impacts of nuclear, biological, or chemical weapons in a China-Taiwan clash.

[5]That this report focuses on military issues should not be interpreted as suggesting that the crux of the China-Taiwan issue is military; neither do the authors believe that military means are the only or even the most likely way of resolving the dispute. Our given task has been to examine the balance of power across the strait, not to document, explicate, or predict the complex political dynamics at the heart of the differences between Taipei and Beijing. We recognize that a strongly deterrent Taiwanese posture is only one part—albeit a vital one—of the equation for maintaining peace and stability on the strait and in East Asia.

sent our recommendations for U.S. policy in Chapter Four along with some brief concluding remarks.

This report includes two appendices. The first contains some thoughts regarding the Chinese missile threat to Taiwan, and the second details the methods and assumptions used in our analytic modeling. A list of references is also attached.

Chapter Two
SCENARIO AND APPROACH

A CHINESE INVASION OF TAIWAN

Although coercive scenarios (e.g., limited missile strikes) are usually regarded as the most likely form of Chinese use of force against Taiwan, we assessed the more extreme case of an outright air and amphibious invasion of the island. We chose to focus on this challenging contingency for six reasons.

First, some analysts argue—despite the common wisdom—that "immediate and full-scale invasion" is the most likely form of conflict between the two sides. One writes:

> Massive surprise attacks have distinguished PLA opening campaigns in the past, such as in Korea in [1950], India in 1962, and Vietnam in 1979. More importantly, [Chinese] military planners believe that the gulf in cross-strait relations would be so wide by the time the leadership resorted to force that limited attacks would be futile in dissuading Taiwan . . . and that the only viable option would be to invade the island. (Cheung, 1997, p. 57.)

Second, as the "worst-case" scenario, it is of interest to military planners whose responsibility it is to deter potential adversaries from dangerous courses of action. Whether China could succeed in invading Taiwan, and under what circumstances, may be an open question. But the enormous political implications and tragic human and economic costs that would ensue should Beijing make the attempt are not.

Third, the possibility of a direct Chinese invasion of Taiwan—and expectations regarding the outcome of such an attack—is important in shaping overall perceptions of the balance between the two sides. Evidence that an invasion appears likely or unlikely to succeed could have an impact on Taiwan's ability to deter any Chinese use of force.

Fourth, while China has other options for using force to coerce or punish Taiwan, the seizure and holding of the island represents a very high-order threat and is the only alternative that guarantees Beijing's control when hostilities end.[1] So, in some sense, the credibility of the invasion threat underwrites the other, lower-level options such as limited missile strikes or maritime harassment. Schelling (1966) notes, "It is the *threat* of damage, or of more damage to come, that can make someone yield or comply. It is *latent* violence that can influence someone's choice."[2] Clearly China could, if it wished, inflict a great deal of damage on Taiwan. If, however, the ROC possesses a robust ability to defeat an invasion attempt, Taiwan could effectively resist forced unification if it were willing to absorb the blows.[3]

Fifth, while it seems unlikely that China would undertake such a desperate gamble, it is important to think through how the PLA might essay the operation and what steps would be needed to defeat it. After all, it was always terribly unlikely that the Soviet Union would launch a massive nuclear attack on the United States. Still, hundreds if not thousands of war games, exercises, and analyses were invested in exploring the "what-ifs" of the contingency, precisely because the consequences of failing to deter it were so dire. While a Chinese invasion of Taiwan would represent a much less dire turn of events than global nuclear holocaust would have been, it is nonetheless a serious enough prospect to warrant at least some attention.

Finally, an invasion scenario incorporates a number of elements that could be components of other coercive strategies directed against

---

[1] There are numerous discussions of alternatives available to China for using force against Taiwan. See, for example, U.S. Department of Defense, 1999; Bitzinger and Gill, 1996; Dreyer, 1999, especially p. 12; and Anderson, 1999.

[2] Emphasis in the original.

[3] If China were willing to unleash its nuclear arsenal on Taiwan, it could almost certainly destroy the ROC as a functioning society and subsequently militarily occupy the rubble. We shall not speculate as to whether such a Pyrrhic triumph would ever appear attractive to the mainland. We shall say only that neither this argument nor our analysis contemplates Chinese use of nuclear weapons against Taiwan.

Taiwan. Although lesser-order conflicts might be more likely to occur, many of them—such as a naval blockade or a protracted low-intensity battle of attrition over the Taiwan Strait—would involve key elements of the cases we examine here. Perhaps the most obvious and significant of these lesser contingencies would be the employment of conventionally armed surface-to-surface missiles (SSMs) against targets in Taiwan.

Since China's 1995 and 1996 "tests," Beijing's arsenal of ballistic missiles has figured prominently as a potential coercive instrument vis à vis Taiwan. It seems likely that any invasion scenario would begin with a barrage of Chinese missiles raining down on key military targets on the island: command and control (C^2) centers, air-defense sites, and air bases. Similar targets would likely be at the heart of any coercive air and missile attacks on Taiwan, at least initially. Therefore, our findings regarding the effectiveness of such strikes in degrading Taiwanese defenses in the context of an invasion would have at least some applicability to the broader question of the military utility of China's missile force.[4]

ANALYTIC STRUCTURE AND METHODOLOGY

Scoping the Problem

Our notional war is set in 2005. Although the Chinese would have a number of options as to the phasing and timing of an attack in that

---

[4]In our analysis, China expends much of its available missile inventory in the attacks on Taiwan, and its front-line aircraft are very heavily committed to the campaign. Some might argue that the PRC would withhold a substantial portion of its forces even from a large-scale attack on Taiwan to ensure that it retained some level of coercive power should the assault fail. We did not consider this strategy for three principal reasons. First, a "fleet in being" strategy is not necessarily a viable alternative for the power that is on the strategic and tactical offensive. As will be seen, our analysis suggests that the PRC has a very hard row to hoe in a full-scale war with Taiwan, even when it commits forces of the size and quality we include. Any reduction in the number of forces Beijing engages would only have worsened these outcomes. Second, most analysts—ourselves included—believe that China would only resort to an all-out attack on Taiwan as a last resort, when all other avenues of influence have been exhausted. In this event, if Taiwan survived a massive attack by China it would, in the aftermath, presumably not be particularly susceptible to more-limited coercive tactics. Finally, the PRC's nuclear capabilities provide it with something of an ultimate trump card in any event.

time frame, analytically we can divide the campaign into four segments.

- In the first phase, the two sides would fight for *air superiority*. Elements of this operation would include Chinese missile and air attacks on ROC air bases, surface-to-air missile (SAM) sites, early warning (EW) radars, and C^2 facilities in addition to air-to-air combat.[5]

- The second phase, which could begin simultaneously with the first, would be a struggle for *maritime control* of the strait, involving air, surface, and submarine forces as well as land-based antiship missile (ASM) units. Elements of antiair warfare (AAW), antisurface warfare, and antisubmarine warfare (ASW) would all be involved.

- Once air superiority was achieved by the mainland, *invasion preparation* would begin. Followup air strikes would focus on destroying coastal strongpoints, destroying ROC artillery and armor concentrations, and generally "softening up" the island's defenses.

- The fourth phase would involve actual *landing operations* on the Taiwanese shore. This phase could include amphibious landings, paratroop assaults, and heliborne attacks directed at gaining a substantial foothold on the island and collapsing Taiwanese resistance.[6]

Our attention is focused mainly on the battle for air superiority and, secondarily, on the contest for control of the seas.[7]

---

[5]We assume that the Taiwanese will concentrate their resources on defense and not launch offensive strikes against air bases on the mainland.

[6]Whether the amphibious assault would be the focal point of the invasion or a supporting operation is a point of some debate but is essentially irrelevant to our analysis. Also, were the Chinese invasion successful, there could be a fifth phase in which the PRC attempts to consolidate its hold on the island, perhaps in the face of determined U.S. attempts to dislodge it. We did not consider any such scenarios in this study.

[7]RAND colleague Michael Swaine notes that even in our base case, we assume that the Chinese military will have made a number of major advances in capabilities; indeed, we will comment on several of the most salient such points as we go along. Our perspective in this study is that of conservative defense planners. Hence, we will tend, where there is uncertainty, to give the Chinese the benefit of the doubt. Some will assuredly argue that we leaned too far in that direction, while others will just as

Historically, it is virtually a truism that control of the air and control of the sea are absolute prerequisites for a successful amphibious or airborne assault. It was the absence of air superiority—which the Luftwaffe had failed to win in the Battle of Britain—that prevented Adolf Hitler from attempting an invasion of Great Britain in 1940–1941. Conversely, it was the Allies' total dominance of air and sea that enabled General Dwight D. Eisenhower to breach Germany's Atlantic Wall in June 1944. The Taiwan Strait is considerably wider than the English Channel and poses a formidable barrier to a potential invader, further strengthening our conviction that no Chinese attack can hope to succeed without first gaining mastery of the airspace above the strait and then of the waters themselves.

This may be particularly true in the context of a PRC attack on Taiwan. The People's Liberation Army Navy (PLAN) owns enough amphibious lift to move about a division of troops at a time, hardly enough to establish and sustain a firm foothold in the face of determined Taiwanese resistance. Therefore, many analyses contemplate a kind of "Dunkirk in reverse," with China employing numerous commercial vessels to transport troops, equipment, and supplies across the strait.[8] Such an operation, involving unarmed merchant shipping, would be sheer folly unless China had secured almost uncontested dominance of the air and sea.[9] Similarly, the kind of large-scale airborne and air assault operations often suggested as part of a PRC attack would be virtually suicidal unless the ROC's air defenses had been thoroughly suppressed.[10]

---

certainly assert the contrary. We believe that we have struck a decent balance between reasonable conservatism and "cloud-cuckoo-land"; we are under no illusions that it is the only such balance.

[8]See, for example, U.S. Secretary of Defense, 1999. A less official but more entertaining depiction of such an operation is Yuan Lin, 1997, translated in Foreign Broadcast Information Service FBIS-CHI-97-268.

[9]Commercial vessels are not compartmented to withstand damage as warships are, nor are their crews trained in the kinds of damage-control procedures that can mean the difference between life and death in a combat situation. It is also worth noting that many commercial bottoms would require an operating port to offload their cargoes. Seizing such a facility intact would present an enormous challenge to the Chinese.

[10]The People's Liberation Army (PLA) currently has only very limited airborne and air-assault capabilities. Each of the three airborne "divisions" is roughly the strength of a U.S. airborne brigade. Given the existing inventory of transport aircraft in the PLA Air Force (PLAAF), it is unlikely that even one "division's" worth of troops could be dropped in a single lift.

Finally, the surface forces of the two navies consist of warships that have very limited air defense capabilities. In the absence of air superiority, the PLAN's warships would be very vulnerable to air attack in the confined waters of the strait.

We therefore conclude that the battle for air superiority in particular is the linchpin of the campaign.

Air War Methodology

For our work, we needed a tool that was sufficiently high-level to permit construction of an open-source database with reasonable effort, while detailed enough to facilitate extensive parametric analyses of the air war. For purposes of credibility, we also wanted a model that had been employed in—and calibrated for—numerous other studies. We chose RAND's Joint Integrated Contingency Model (JICM) as best fitting these criteria. JICM is a theater combat model designed to support the kind of exploratory analysis that we emphasized in this project.[11] After preparing a database from open-source materials, we conducted more than 1,700 model runs to examine both a baseline scenario and numerous what-ifs.[12]

We made an initial set of model runs to identify the factors that seemed likely to play a determining role in the outcome of the war over the strait. We then conducted extensive sensitivity analyses on seven variables:

- The size and composition of the air forces committed to the attack by the PRC.
- Each side's possession of beyond-visual-range (BVR), "fire-and-forget" medium-range air-to-air missiles (AAMs).

---

[11] For a full description of JICM, see Jones and Fox (1999).

[12] Among the sources used were: International Institute of Strategic Studies, 1998; U.S. Naval Institute, 1999; Taylor, 1988; Wang, 1999; Jane's Information Group, 1998; Sharpe, 1998; Cullen and Foss, 1997; Jackson, 1998; *World Navies Today*, 1998–1999; U.S. Secretary of Defense, 1999; and various issues of the following journals: *Aviation Week & Space Technology, International Defense Review, Jane's Defence Weekly,* and *Jane's Intelligence Update.*

- The number and quality of short- and medium-range ballistic missiles (SRBMs and MRBMs) used by the Chinese.

- The number of advanced precision-guided munitions (PGMs), such as laser-guided bombs (LGBs) and Global Positioning System (GPS)-guided weapons in the Chinese inventory.

- The ability of the Republic of China Air Force (ROCAF) to generate combat sorties.

- The quality of the ROCAF's aircrew.

- The extent, if any, of U.S. air forces, both land and sea based, committed to Taiwan's defense.

We will briefly discuss each in turn.[13]

PRC Force Size and Composition. Significant uncertainty surrounds the number of air forces the PLAAF would commit to a struggle with Taiwan. Only a limited number of bases are available within operating distance of the strait, and the PLAAF has virtually no capabilities for midair refueling of fighter aircraft. Also, the airspace in and around Taiwan is very limited, which would restrict the number of aircraft that either side could commit to the fight at any one time. Finally, the PLAAF has had little experience with the management of large groups of aircraft and would likely experience serious C^2 difficulties in a complex, swirling air battle.

To reflect this uncertainty, we used two differently sized Chinese air forces in our analysis, as shown in Table 2.1. The base case reflects our best estimate of the number of aircraft that could be operated from the existing array of PLAAF bases in the vicinity of Taiwan.[14] Note that this force includes the bulk of China's most modern fighters, such as the Su-27.

---

[13]A more complete discussion of the JICM representation of the China-Taiwan air war may be found in Appendix B to this report.

[14]Our assessment is that the PLAAF would have to become much more skillful in employing large groups of aircraft to take maximum advantage of a force of this size. Although it is highly unlikely that such progress could be made by 2005, we believe that principles of conservative planning, from the defender's point of view, analytically justify our assumption that forces of this size could be used.

Table 2.1

PLAAF Forces Committed to Taiwan Contingency

Type	Base Case	Big Force	Advanced Force
Su-27 Flanker	72	72	144
Q-5 Fantan	120	216	0
JH-7	48	48	72
J-7 Fishbed	168	288	144
J-8 Finback	144	288	144
J-10	24	24	48
Su-30 Flanker	0	0	24
H-6 Badger	48	48	48
AWACS	6	6	6
Miscellaneous	49	49	49
Totals	679	1,039	679

SOURCE: Order of battle from the International Institute for Strategic Studies, 1998, p. 180, and the authors' projections.

The "big" force is half again as large as the base case and is meant to represent a strategy whereby the PLAAF forward deploys additional aircraft to take the places of those lost in action. In our analysis, enough of these "attrition replacements" flowed forward on a daily basis to keep 679 jets in action up to the limit of 1,039 total aircraft committed.[15]

A second uncertainty we wanted to capture concerns the pace of PLAAF modernization. To reflect this, we created the "advanced" force shown in the table. The same size as the base case, it contains more than twice as many fourth-generation fighters (216 versus 96).

AA-12 and AMRAAM. Both China and Taiwan have been actively pursuing the acquisition of modern air-to-air weapons, with China seeking to buy the Russian AA-12/R-77 Adder and Taiwan negotiating for the U.S. AIM-120 Advanced Medium-Range Air-to-Air Missile (AMRAAM).[16] While neither side currently fields such a weapon, it is

---

[15]This may also overstate Chinese capabilities. We have seen no evidence that the PLAAF has extensively rehearsed rapidly deploying air force squadrons from one base to another, for example.

[16]In April 2000, the U.S. announced that it would sell AMRAAM to Taiwan but keep the weapons stored in the U.S. until China fielded a comparable capability. Most analysts

certainly possible that one or both will by 2005.[17] We therefore analyzed four cases of BVR capability: one in which neither side possesses them; a second, in which Taiwan has AMRAAMs (carried by its fleet of F-16s); a third, in which the PLAAF has AA-12s (carried by its Su-27s, J-10s, and, in the advanced case, Su-30s); and the fourth, in which both sides are so equipped. Air forces with BVR weapons were given sufficient stockpiles to last for four days of intensive combat.

China's Missile Force. We explored the potential impact of ballistic missiles on the campaign by postulating two different missile forces for the Chinese. Table 2.2 lists the number of missiles we made available for use in the base case. These are weapons with either advanced unitary high-explosive or cluster munition warheads. Half of the DF-15s also employ GPS-aided guidance (as noted) to increase their accuracy. Half of our cases used these 310 missiles. In the others, we doubled the size of the PRC's missile force to 620 missiles.

Table 2.2

Chinese Missile Forces

Missile Type	Quantity	Range (km)
DF-3	20	2,800
DF-21	80	1,800
DF-11	50	280
DF-15	80	600
DF-15 with GPS	80	600

SOURCE: Lennox, 1999, and authors' projections.

---

believe that, regardless of this move, China will persist in attempting to acquire and deploy the AA-12.

[17]Taiwan does have a number of French MICA missiles deployed on its fleet of Mirage 2000 fighters. The MICA can be fitted with an active radar seeker that, combined with an inertial navigation system, guides it autonomously to targets at short ranges. However, maximum-range launches—the kind preferred by fighter pilots—require the missile to receive targeting updates from the Mirage radar until the active radar can lock on and track the bogey. So, while the MICA is the most advanced AAM currently available to either side, it does not completely fit the description of a true "fire-and-forget" AAM.

We assumed that Taiwan would deploy no effective active missile defenses by 2005.[18]

Chinese PGM Inventories. Another uncertainty regards the number and quality of air-delivered PGMs, such as LGBs and satellite-guided munitions, available to the PLAAF in 2005. In half of our runs, the Chinese employ a very limited supply of about 300 PGMs, a quantity broadly consistent with the very limited capabilities possessed by today's PLAAF. The other half of our cases featured a much larger Chinese stockpile of 3,000 PGMs for use against Taiwanese targets.

ROCAF Sortie Generation. The JICM calculated degradation to sortie generation capabilities resulting from Chinese air and missile attacks on Taiwanese air bases. In case of war, these installations are likely to also be the targets of attacks by Chinese special operations forces (SOF), which could further impede flight operations. Also, the ROCAF has never been called on to maintain a very high tempo under wartime conditions, and we wished to explore the impact of any inability on its part to sustain such intense activity. We therefore used three different levels of ROCAF sortie rates: 100 percent of baseline, 75 percent, and 50 percent.[19]

ROCAF Pilot Quality. Pilot training is a key variable in air combat. Our base case assumption, deriving from unclassified estimates of flying hours and conversations with experts both in the United States and Taiwan, is that a ROCAF pilot is about 80 percent as well-trained as his U.S. counterpart, while a PLAAF flyer is only about half as good as the American. To see what effect pilot skill might have on combat outcomes, in half the cases, we more pessimistically rated ROCAF aircrew as only 60 percent as skillful as U.S. flyers.

---

[18]It is possible that Taiwan could begin deployment of PAC-3 surface-to-air missiles with improved antimissile capabilities by 2005, but it seems to us doubtful that it could have them operational in sufficient numbers to greatly affect the outcome of the sorts of massive attacks China employs in this analysis.

[19]In this case "baseline" refers to the sortie generation potential of a base *before* taking into account any damage from Chinese air and missile attacks. Assume a base is capable of generating 200 sorties per day when undamaged and operating at full efficiency. It would produce 200, 150, and 100 sorties each day at the 100, 75, and 50 percent levels of efficiency. If it had also absorbed 10 percent damage from Chinese strikes, these values would be reduced to 180, 135, and 90, respectively.

U.S. Forces Engaged. Even if we assume that the United States would directly assist Taiwan in its defense against invasion, it is unclear how much force would be brought to bear quickly—during the crucial four-day period covered by our simulations. To capture this uncertainty, we used six different levels of direct U.S. combat involvement:

- No U.S. forces engaged.
- A single CVBG operating east of Taiwan.
- A single wing of USAF 72 F-15C fighters stationed at Kadena AB on Okinawa.
- One CVBG and one F-15 wing.
- Two CVBGs.
- Two CVBGs and one F-15 wing.

Operationally, we assumed that the carrier, operating close to the war zone, could surge additional sorties to meet incoming Chinese attacks. The USAF fighters are based so far away, however, that they could not be so responsive; they were assumed to maintain combat air patrol orbits.[20]

Combining all the possible permutations of these factors (three PLAAF force structures, four BVR cases, two levels of Chinese SSM, two levels of PLAAF PGM stocks, three levels of ROCAF sortie generation, two levels of ROCAF crew training, and six levels of U.S. involvement) yields our 1,728 cases, as shown in Table 2.3.[21]

---

[20]Twenty-five percent of all U.S. fighter sorties, both land- and carrier-based, were withheld for local air defense and other unmodeled missions.

[21]Note that each case represents a unique configuration of these seven parameters. Therefore, results across cases do not represent outcome "probabilities" in the sense that they would in, for example, a Monte Carlo analysis. Neither should a small percentage of bases be interpreted as a reason for complacency. The analysis focuses on variables that are neither random nor truly independent; they represent instead the results of policy choices by the actors involved, and several of them are under the control of the Chinese. The results therefore should be read as identifying situations—less well-trained ROC pilots confronting an advanced PRC threat without U.S. assistance, for example—that appear to bode more or less well for Taiwan's defensive prospects. The bigger the percentage of bad cases, the more such situations there appear to be, and the weaker we would judge Taiwanese defensive capabilities.

Table 2.3

Cases for Exploratory Analysis

Variable	Values Used
PLAAF force	Base, big, advanced
Advanced BVR capabilities	None, Taiwan only, PRC only, both
PRC missile force	310, 620
PLAAF PGMs used	300, 3,000
ROCAF base sortie rate	100%, 75%, 50%
ROCAF pilot quality	80%, 60%
U.S. forces engaged	None, CVBG, FW, CVBG plus FW, 2 CVBGs, 2 CVBGs plus FW

Naval War Methodology

Our more limited analysis of the naval war was undertaken using both the JICM and *Harpoon*, a computer-based simulation of maritime warfare. *Harpoon* is widely considered the best commercially available depiction of modern maritime combat. It includes representations of submarine, surface, and air warfare.

This study also benefited from numerous discussions with area experts, analysts, and military officers in the United States and overseas. Particularly useful were the insights gathered on a trip to Taiwan, hosted by the ROC Ministry of National Defense (MND). The information gained in these exchanges helped shape the inputs to the analysis and was critical in helping us understand its results.

Caveats

This work explores only a very limited region of what is often referred to as the "scenario space." We concentrated on one specific scenario involving one particular Chinese offensive strategy, and we selected the factors to vary based on our reading of the extant literature on the China-Taiwan balance as well as discussions with experts in the United States and elsewhere. We also focused our attention on what might be thought of as "reasonable" cases: those reflecting current capabilities, linear projections of current capabilities, and capabilities conceivably attainable within our limited time frame. To do otherwise would have required a level of effort well beyond the scope of this project and one inconsistent with its goal of providing insights into U.S. options for improving Taiwan's self-defense capabilities.

As such, we present these results as *illustrative* and *indicative*, meant to highlight and illuminate certain key points that emerged from our overall analysis. We ask the reader to bear this in mind throughout this report.

ORDERS OF BATTLE

Because five years is not a particularly long span of time in terms of military capabilities, much of the two sides' respective arsenals in our scenario resides already in their inventories. So, the orders of battle we used consist mainly of familiar systems.

Air, Air Defense, and Missile Forces

Table 2.4 shows the composition of the ROCAF used in the analysis. It includes almost 350 modern combat aircraft (plus eight E-2Ts and 44 miscellaneous aircraft).[22] As noted earlier, we varied the size and composition of the committed fraction of the PLAAF (as shown in

Table 2.4

ROCAF Composition

Type	Quantity
F-16A/B	162
Mirage 2000	54
Ching Kuo IDF	126
E-2T AWACS	8
Miscellaneous	44
Total	394

SOURCE: Order of battle from International Institute for Strategic Studies, 1998, p. 198.

---

[22] As of spring 2000, the ROCAF is experiencing continued difficulties integrating the F-16 into its force structure; the entire fleet has been grounded on several occasions. We assume that these teething problems will have been overcome by 2005.

Table 2.1) to reflect uncertainties in both the proportion of the PRC's forces that would be engaged in an attack on Taiwan and the results of Beijing's modernization efforts.

Table 2.5 lists the surface-to-air order of battle for Taiwan's integrated air defense system (IADS).[23] In the event of war, the radars and C^2 elements for these systems would likely be high-priority targets for Chinese missiles and SOF. We have already discussed the numbers and kinds of SSMs employed by China in the scenario (Table 2.2 above).

Naval Forces

Tables 2.6 and 2.7 list the naval forces we used for each combatant.[24] As with the air forces, the table shows only that portion of the PLAN that we assumed would be committed to the attack on Taiwan. Again, as with the air force, it includes the bulk of the navy's modern combatants.

The tables show the ROCN to be outnumbered in terms of surface warships by 65 to 37. However, by and large, Taiwan's navy holds a significant qualitative edge over the PLAN:

Table 2.5

Taiwan Surface-to-Air Order of Battle

Type	Number of Batteries
Patriot PAC-2	9 (6 quad launchers each)
Improved Hawk	36 (18 triple launchers each)
Tien Kung	6 (6 quad launchers each)

SOURCE: Inventory figures extrapolated from U.S. Naval Institute, 1999.

---

[23]The Tien Kung is a Taiwanese-produced Patriot-like SAM. We did not model Taiwan's numerous low-level air defense systems, such as Chaparral and Avenger.

[24]The size of the PLAN's submarine fleet, and particularly the number of Kilos it could field by 2005, appears to be exaggerated in our order of battle. However, our assessment is that the results we saw in our naval combat simulations would not be particularly sensitive to the precise number of Chinese subs engaged; in any event the PLAN would field far too many for the ROC Navy (ROCN to effectively cope with.

Table 2.6
Taiwanese Naval Order of Battle

Class	Type	Qty	SSM	SAM
Chien Yang	DDG	7	4 x Hsiung Feng II	10 x SM-1R
Fu Yang	DD	2	5 x Hsiung Feng I	4 x Sea Chaparral
Po Yang	DD	1	5 x Hsiung Feng I	—
Kun Yang	DD	3	5 x Hsiung Feng I	4 x Sea Chaparral
Cheng Kung	FFG	8	8 x Hsiung II	40 x SM-1R
Kang Ting	FFG	6	8 x Hsiung II	4 x Sea Chaparral
Chin Yang	FFG	10	4 x Harpoon	—
Hai Lung	SS	2	—	—

SOURCE: IISS, 1998; Sharpe, 1998; USNI, 1999.

Table 2.7
Chinese Naval Order of Battle

Class	Type	Qty	SSM	SAM
Sovremenny	DDG	2	8 x SS-N-22	48 x SA-N-7
Luhai	DDG	2	16 x C-802	8 x Croatale
Luhu	DDG	2	8 x C-802	8 x Croatale
Luda III	DD	1	8 x C-801	—
Luda II	DDG	11	6 x C-201	8 x Croatale
Luda I	DDG	14	6 x C-201	8 x Croatale
Jiangwei	FFG	6	6 x C-802	6 x HQ-61
Jianghu	FF	27	4 x C-201	—
Han	SSN	2	—	—
Type 93	SSN	2	—	—
Song	SS	3	—	—
Kilo	SS	10	—	—
Ming	SS	4	—	—
Romeo	SS	3	—	—

SOURCE: IISS, 1998; Sharpe, 1998; USNI, 1999.

- Taiwan's 14 *Cheng Kung* and *Kang Ting* frigates—modified versions of the U.S. *Oliver Hazard Perry* and French *Lafayette* classes, respectively—are probably the most modern and well-balanced combatants available to either side.

- The PLAN's two Russian-built *Sovremennys* carry the very dangerous SS-N-22 "Sunburn" sea-skimming antiship missile. However, it appears that the Chinese will only be acquiring two hulls of this class, whose AAW and ASW capabilities are less impressive than those fitted to both the *Cheng Kung* or *Kang Ting*.²⁵

- In terms of antiship missile mounts, the PLAN holds a 366 to 210 edge. Almost half of the Chinese missiles are obsolescent C-201s; however, this missile is large, slow, and vulnerable to countermeasures and such close-in weapons systems as the Phalanx guns that equip nearly all ROCN combatants. Most of the ROCN missiles, meanwhile, are more-modern types, such as the Hsiung Feng II, which is broadly similar to the U.S. Harpoon.

- The *Cheng Kung* frigates along with the aging *Chien Yang* destroyers (former U.S. *Gearing* class) are equipped with the Standard SM-1 medium-range air defense system, which has significantly longer range and more capability than any SAM in the PLAN's arsenal.

If the surface forces appear evenly matched, it is quite a different story beneath the waves. Beijing has applied enormous political pressure to prevent foreign suppliers from selling attack submarines (SS) to Taiwan, which has no indigenous submarine production capacity. So, the ROCN's two *Hai Lung* boats (Dutch *Zwaardvis* class) are left to confront a much larger number (24 in our scenario) of PLAN submarines. Although many of the Chinese submarines are older and less capable, the *Kilo* and *Song* classes are fairly advanced, and the Type 93 SSN is expected to be similar to the Russian *Victor II* class in performance. Although not nearly in the class of the very latest U.S. or Russian SSNs, the Type 93 will nonetheless be by far the most capable sub ever deployed by the PRC. Combined with the acoustic qualities of the Taiwan Strait, which make it a nightmare for ASW operators, this disparity could have a telling impact on the battle for maritime control.

---

²⁵In the long term, the Chinese may be able to adapt other surface combatants to fire the SS-N-22. However, we think it highly unlikely that they will be able to do so by 2005. Also, there are reports that Moscow has agreed to sell China two more ex-Russian Navy *Sovremennys*. See Novichkov, 2000.

Command and Control

The analysis required many assumptions and the problem frequently arose as to how much credit to give the protagonists for various capabilities. We decided to credit both sides with taking measures to increase their competence in critical areas. Because of these assumptions, our analysis is less a current net assessment of actual capabilities on the two sides than an assessment of reasonable *potential* capabilities with given orders of battle.

In particular, we credited the Chinese with more capability than they have demonstrated in conducting complex offensive operations. PLAAF training and exercises have not typically featured large numbers of aircraft engaged in coordinated activities, and there have been no recorded joint exercises that even approached the complexity of a Taiwan invasion scenario. Hence, we are probably being rather conservative (from a Taiwanese planner's perspective) in crediting the mainland with the ability to execute such an intricately choreographed air and missile operation.[26]

We also assumed that Taiwan would be able to maintain the basic functionality of its C^2 system even under the stress of a concerted PRC attack.[27] To the extent that it relies on fixed surveillance radars and unhardened command posts, the ROC's new "Strong Net" air-defense system will continue to be vulnerable to air, missile, and SOF attacks.

That said, given the small size of the battlespace and the large number of forces engaged, it seems reasonable to assume that each side's combat jets and warships could "find the fight." As explained in Appendix B, we constrained both sides' aircraft in a manner we believe to be consistent with their likely performance in a "target-rich" combat environment under conditions of imperfect C^2.

---

[26]The PLAAF last engaged in air-to-air combat during the 1958 Taiwan Strait crisis. The results were not encouraging for Beijing: the Chinese were unable to gain air superiority over the strait and, according to USAF statistics, suffered 32 combat losses to only three for the ROCAF. See Allen, Krumel, and Pollack, 1995, pp. 61–69.

[27]Including possible, but unmodeled, information warfare operations.

PLAYING OUT THE SCENARIO

Overview

Our analysis suggests that any near-term Chinese attempt to invade Taiwan will likely be a very bloody affair with a significant probability of failure. Leaving aside potentially crippling shortcomings that we assumed away—such as logistics and C^2 deficiencies that could derail an operation as complex as a "triphibious"[28] attack on Taiwan—the PLA cannot be confident of its ability to win the air-to-air war, and its ships lack adequate antiair and antimissile defenses. Provided the ROC can keep its air bases operating under attack—a key proviso we discuss at length in Chapter Three—it stands a good chance of denying Beijing the air and sea superiority needed to transport a significant number of troops safely across the strait.

The War in the Air

Although we varied the particulars of the Chinese strategy across runs, the air battle was laid out roughly as follows:

- An initial barrage of Chinese tactical ballistic missiles (TBMs) aimed at early warning radars, SAM sites, and airfields.

- A wave of Chinese fighters performing a sweep over the strait and the island.

- A large, escorted strike package going after coastal defense, air defense, and airfield targets.

- A second fighter sweep approximately four hours later.

- A third sweep followed by a second wave of strikes four hours after that.[29]

This pattern was repeated for four days.[30]

---

[28]Amphibious, airborne, and air assault.

[29]Given the known capabilities of the two sides, we constrained the air war to daylight only.

[30]The four-day campaign length was arrived at experimentally—our preliminary model runs indicated that the air battle tended to resolve itself one way or another by that time. This should not be interpreted as suggesting that one or the other side is

We scored the air war in each of our model runs using the PRC-ROC exchange ratio—the number of mainland Chinese aircraft shot down for every Taiwanese plane lost—as the measure of merit. While an imperfect indicator, it does provide some feel for the overall direction and dynamics of the air battle.

Each outcome was evaluated based on the opening ratio of the two sides' committed air forces (679:394, or 1.72:1 in the base and advanced variants, 1,039:394, or 2.63:1 for the big threat case) and the total losses after four days of combat. Cases in which the ROC achieved an exchange ratio 50 percent greater than the opening force ratio (2.58:1 in the base and advanced cases and 3.95:1 in the big cases) were scored as "green" (which appears as black in the figures), meaning that Taiwanese forces could almost certainly deny the PRC a viable invasion opportunity.[31] When the final exchange ratio was less than the "green" threshold but greater than the opening force ratio (1.72 or 2.63), we counted the outcome as "yellow" (medium gray), meaning that the ROC was at least holding its own and could probably deter or defeat a landing attempt. Any case in which the exchange ratio dropped below the "yellow" threshold was counted as "red" (light gray), meaning that the PRC could plausibly mount an invasion attempt.[32]

Figure 2.1 shows how all the model runs scored out, broken out by the size and quality of the PRC air forces brought to bear.[33] Almost 90 percent of the trials against the base PRC threat resulted in Taiwanese "victories"—that is, the outcome of the air war seemed likely to prevent a successful invasion attempt. Even when China was permitted to bring more of its airpower to bear, about 75 percent of

---

likely to give up after a few days of pitched fighting. Indeed, if Taiwan were left to stand alone against the mainland, China would stand a relatively good chance of grinding down the ROC's defenses in a protracted war of attrition. Our work does not address that potential course of events.

[31]As RAND colleague Paul Davis reminds us, this calculation generates what is known as a "ratio of fractional loss rates," which "determines who wins the battle in a deterministic drawdown" according to the Lanchester square law.

[32]Note that we do not mean that a "red" outcome means that a Chinese invasion would *succeed*; only that the results of the air battle would not necessarily preclude a viable attempt.

[33]"All runs" include those both with and without direct U.S. involvement. We will dissect the effects of that and other key factors here and in the next chapter.

26 Dire Strait?

the outcomes favored Taiwan. However, results were very different when the scenario pitted Taiwan against the more modernized "advanced" mainland force. Only about half of these runs resulted in defense-favorable outcomes.[34]

To make sensible recommendations for improving Taiwan's defensive posture, we need to understand what drives the "bad" outcomes. Table 2.8 shows how our six experimental parameters contributed to each of the "red" cases reported in Figure 2.1.[35]

Against the most likely PRC force, the base case, we see that limited U.S. involvement[36] is a factor in *every* bad case, and both Red BVR

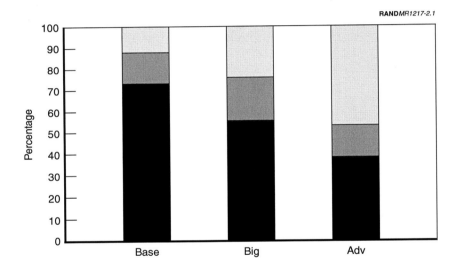

Figure 2.1—Overall Outcomes

---

[34]Of all the cases we ran, 24 covered what many analysts might consider to be the most likely near-term situation: the base PRC threat, no direct U.S. involvement, and no advanced BVR capabilities on either side. Of these results, 17 (71 percent) were coded "green," six (25 percent) were "yellow," and only one (4 percent) came out "red."

[35]The seventh parameter, the size and quality of the PLAAF force employed against Taiwan, is accounted for in the breakout of the cases in the figure.

[36]Defined here as no U.S. forces, one CVBG, or one fighter wing.

superiority and suppressed ROC sortie generation play a role in 80 percent of the "red" outcomes. Put differently, we ran 288 simulations against the base threat in which "BVR parity" was maintained (neither or both sides having AA-12/AMRAAM); of these, only 13 (less than 5 percent) resulted in "red" outcomes. Of 192 cases in which the ROC was able to fly sorties unimpeded against the base threat, only 13 (about 7 percent) results were coded as bad for Taiwan.[37]

Against the big threat, limited U.S. involvement remains the best predictor of a bad outcome, and BVR parity is still critical as well, with less than 15 percent of cases in which either both or neither side had advanced AAMs producing "red" results. Less than 20 percent of the cases in which Taiwan had full sortie generation capability wound up in the "red" category.

As might be expected, the advanced threat produced the most diffuse results; the overall quality of the PLAAF force projected here makes it a much tougher adversary even absent any degradations being imposed on Taiwan. Nevertheless, we see that suppressed sortie generation and limited U.S. involvement are still implicated in more than 60 percent of the poor results.

All cases were characterized by extraordinarily high attrition rates, revealing an air war of great intensity and unprecedented attrition. The median loss rates for the two sides after four days of combat

Table 2.8

Impact of Parameters on "Red" Outcomes

Parameter	Base Case	Big Threat	Advanced Threat
Limited U.S. involvement	100%	81%	62%
PRC BVR superiority	80%	71%	51%
ROC sorties suppressed	80%	75%	64%
Poor ROC training	67%	65%	54%
More PRC TBMs	56%	52%	51%
More PRC PGMs	56%	54%	48%

---

[37]Remember that PRC missile and air attacks had an effect on Taiwanese sortie generation distinct from the exogenous reductions we imposed as an analytical parameter.

were about 75 and 45 percent for the PRC and ROC, respectively. With Chinese missiles and air strikes suppressing a significant proportion of Taiwan's land-based air defenses, the overwhelming majority of these kills came in air-to-air encounters.

Over the four days, the Chinese sustained on average about a 30 percent per sortie attrition rate, while the ROCAF absorbed about 15 percent per sortie. Both numbers are extremely high by historical standards.[38] On the second day of the 1973 Arab-Israeli war, for example, the Israeli Air Force (IAF) lost 22 aircraft in 488 sorties, a 4.5 percent loss rate. In the war as a whole—widely regarded as a catastrophe for the IAF—the Israelis lost 108 aircraft in 7,290 sorties, about a 1.5 percent attrition rate, a factor of 20 smaller than that inflicted on the Chinese in our simulation (Nordeen, 1990, p. 146).

Could these astonishing loss rates represent a reasonable characterization of a near-term China-Taiwan air battle? Perhaps the most instructive historical analog to the China-Taiwan clash we are analyzing would be the Battle of Britain. That campaign, like this one, was intended to pave the way for an invasion. Also like our fictional war over the strait, the Battle of Britain was largely fought in a confined space, over the English Channel and southeastern England.[39] Finally, the general flow of the two battles are similar: the Germans and Chinese each launched large raids of bombers and fighters against which the defending side would mass as strong a challenge as it could.

Even the Battle of Britain, however, did not see losses of the scale we portray in the China-Taiwan air war. The most intense day of fighting in the Battle of Britain was probably August 15, 1940, when the Luftwaffe flew 1,786 sorties. Although losses on both sides were heavy—the British lost 35 aircraft and the Germans 76—the Luftwaffe's per-sortie attrition rate amounted only to 4.2 percent (Terraine, 1985, pp. 186–187).

---

[38]The median figures are quite close to the means: 12 percent for the ROC and 30 percent for the PLAAF.

[39]Taiwan itself covers less than 36,000 square kilometers, which is a little smaller than Maryland and Delaware combined. Roughly double this figure to cover almost all of the strait, and the resulting area is about halfway between West Virginia and South Carolina in size (U.S. Central Intelligence Agency, 1999).

That having been said, it should be noted that there have been examples of extremely heavy attrition in air warfare. On August 17, 1943, 315 bombers from the U.S. Eighth Air Force (8th AF) set out from bases in England to attack the German ball-bearing factory at Schweinfurt; 60 (19 percent) did not return. In October of that year, the 8th AF flew 1,200 sorties in four raids over a period of seven days; 148 bombers were lost, a 12.4 percent attrition rate per sortie (McFarland and Newton, 1991, pp. 127–129). This number is beginning to look more comparable to those racked up in our analysis.

The Battle of Britain analogy may yet prove useful. Consider that battle, refought with modern sensors, aircraft, and weapons. In 1940, radar was in its infancy, and none of the aircraft in the battle had onboard sensors other than the pilot's eyes. The main armament on each side were the .30-caliber and 7.62-mm machine guns, firing unguided projectiles with an effective range measured in hundreds of yards. There were no airborne warning and control (AWACS) aircraft and no data links to permit efficient target allocation. The Germans typically flew a single daily raid, while the Chinese in our campaign are mounting five waves *each day*. Clearly, a modern rendition of the Battle of Britain—with hundreds of radar-equipped jets engaging one another with long-range AAMs under the direction of both airborne and ground-based controllers—would very likely be a much more lethal environment than was the original.[40] In this context, loss rates several times those recorded at the height of the struggle between the RAF and Luftwaffe may be quite credible.[41]

---

[40]The radar in the ROCAF Ching Kuo Indigenous Defense fighter (IDF) is credited with a search range of about 150 kilometers. Assuming it scans a 90-degree arc in front of the aircraft, a single aircraft can track targets over an area of about 18,000 square kilometers, or nearly one-quarter of the entire aerial battlefield. Given limitations on vertical scan, the proportion of the total volume surveyed would be significantly less, but each IDF pilot would still be looking at a considerable wedge of the total battle space. IDF data from Jackson (1998, p. 486).

[41]To the extent that our analysis overestimates Chinese capabilities—which it almost certainly does—it likely also overestimates both sides' losses. To a lesser degree, we may also be giving the ROCAF more credit than it deserves, particularly for being able to sustain operations under attack. This too would result in exaggerated kill rates. Thus, we would not necessarily expect to see such extraordinary losses racked up in a real-world showdown between China and Taiwan. Nevertheless, for the reasons we describe, we would expect the air war between the two sides to feature much greater attrition, in terms of loss rates, than have been seen in perhaps any other air war.

The War at Sea

As our work progressed, it became clearer and clearer that the air war held the essential key to the scenario. Further, the analysis suggested that Taiwan could, if proper steps were taken, have a reasonable degree of confidence in its ability to defeat a Chinese air offensive and thereby prevent a successful invasion. So, our weight of effort gradually shifted to focus increasingly on understanding the air campaign, with commensurately less attention being invested in the naval war. Nonetheless, a few points of interest are worthy of note.

As in the air war, the naval contest in the strait would be very, very bloody. Each side's navy has substantial weaknesses—the ROCN's lack of submarines and the PLAN's limited air defense capabilities—that quickly turn the constricted waters of the Taiwan Strait into a warship graveyard. Our analysis is not sufficiently exacting to support specific attrition estimates, but if the two navies were to meet head-on in the strait, neither could expect an easy victory, and the Chinese certainly cannot be confident of winning.

Our base case pitted the ROCN against the main battle units of the PLAN in an initial battle beginning on D day. After this has run its course, the invasion fleet, consisting of lightly escorted assault task groups screened by numerous small combatants and missile craft, begins to move across the strait. In this variant, PLAN submarines took a heavy toll of ROC surface combatants. As noted above, the strait is a terrible ASW environment. It is doubtful that Taiwan's fixed- and rotary-wing ASW aircraft will be able to operate effectively in the midst of the air battle roaring around them. And, even the ROCN's better ASW surface platforms, such as the modern *Cheng Kung* and *Kang Ding* frigates, may be overwhelmed by the combined air, surface, and subsurface threats.

Another major problem facing both sides will be targeting. Airborne surveillance platforms will be a prime target for the other side's antiair operations, and both sides will likely employ heavy jamming and other electronic warfare techniques. Hence, weapons may not be employed at maximum standoff range, and fratricide will certainly be a concern. The side best able to keep a clear and coherent picture of the evolving battle will stand the better chance of prevailing.

Chapter Three

ISSUES AND IMPLICATIONS

In this chapter, we discuss seven key points that have emerged from our analysis. They have been grouped into two broad categories, one dealing with air superiority and the other with the war at sea.

AIR SUPERIORITY

Base Operability

Even in our base case, the PLAAF outnumbers the ROCAF by 1.7 to 1; the "large" Chinese force enjoys an even greater quantitative edge. This makes it imperative for the Taiwanese to make the most of their smaller force structure by sustaining their ability to generate sorties over time. Equally, the PRC will undoubtedly attempt to disrupt air base operations with missile and air strikes.

Other RAND analyses suggest that attacks employing sufficient numbers of advanced conventional warheads could in fact cause significant damage to unhardened facilities, aircraft parked in the open, and personnel (Stillion and Orletsky, 1999). Figure 3.1, based on our work, suggests the possible implications. The figure shows results at three levels of ROCAF sortie generation capability: 50, 75, and 100 percent *when no U.S. forces are engaged*.[1] Taiwan achieved successful outcomes (green or yellow—black and medium gray in the figures) in about 60 percent of the cases when its air force was flying

---

[1] We removed U.S. forces from the equation to clarify the impact of the changes in Taiwanese sortie rates.

32 Dire Strait?

at full tempo, but only about 40 percent of the time when its sortie generation was reduced by half.[2]

We therefore conclude that ensuring that their air bases remain operational under wartime stresses must be a primary worry for Taiwanese planners. The following four vulnerabilities stand out in our assessment:

- Aboveground fuel storage tanks and parking facilities for tanker trucks would make appealing targets for missile and air attacks. Any curtailment of aviation fuel supply would have an obvious and deleterious effect on the ROCAF's operations tempo, an effect that could be fatal in case of a large-scale Chinese assault on the island.

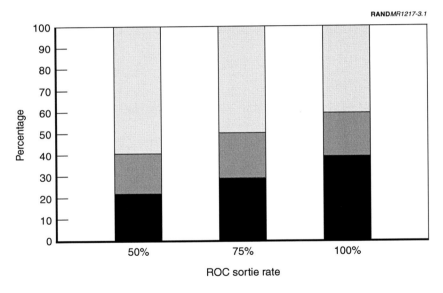

Figure 3.1—Effects of Reductions in ROCAF Sortie Rates

---

[2]Recall that even in the "100 percent" case, Chinese air and missile attacks on air bases can have an impact on the ROCAF's sortie rates. The reductions described here are in addition to these effects.

- Like fuel tank farms, large, unhardened engine and avionics support shops could be put out of action by air- or missile-delivered ordnance or by SOF action.

- Operating surfaces—runways and taxiways—could be potholed with submunitions, and repair operations could be impeded by scatterable mines. Even brief closures will be difficult to tolerate in a pitched battle for air control.

- Finally, although most experts discount the possibility of Chinese weapons of mass destruction (WMD) use against Taiwan, it does not seem unreasonable that the PLA might employ chemical agents against large military targets, such as air bases.

The ROCAF is not ignorant of these threats. It has built at least two large bases buried within mountainsides. Whether future Chinese PGMs will be capable of destroying or blocking the doors into and out of these facilities is unclear, as is the ROCAF's ability to conduct all aspects of combat base operations within the confines of these bases. In the near term, however, it is likely that some aircraft and personnel will at least be safe from nonnuclear attack while inside these redoubts.

The ROCAF also plans to rely on dispersed operations, including flying aircraft into and out of numerous highway strips scattered across the island. This strategy, while increasing survivability against certain kinds of threats, carries significant costs as well. Dispersal greatly complicates logistics, for example, and requires duplication of maintenance manpower and equipment across all the alternate operating sites. It also may significantly increase the dispersed unit's vulnerability to SOF attacks.[3]

In many regards, the ROCAF's bases face a situation similar to that confronted by NATO's air forces in the Central Region of Europe in the 1970s and 1980s. Like the Taiwanese, NATO strategists faced an

---

[3] In this, the ROCAF high command may face the same dilemma that confronted U.S. Army Air Forces Commander Lieutenant General Walter C. Short at Pearl Harbor in early December 1941. Forewarned of possible hostilities with Japan, he chose to gather the aircraft under his charge from their dispersed locations to make them easier to protect against enemy saboteurs. In so doing, he created a nice, concentrated target for the fighters and bombers of the Imperial Japanese Navy on December 7. For more on the difficulties inherent in dispersed operations, see Stillion and Orletsky (1999).

adversary who outnumbered them and to whom the first move was conceded. That first move was expected to include massive air, missile, and SOF attacks on a number of targets in NATO's rear area, including the alliance's air bases.

To counter this, NATO made enormous investments in passive defenses to ensure air base operability. Fuel storage was buried and pipelines were laid so aircraft could be fueled from hydrants in the floors of their shelters rather than by vulnerable tanker trucks. Maintenance facilities were hardened. Engineering equipment was procured, and stockpiles of runway-repair materials were laid in. Ground and aircrew regularly practiced performing their duties while wearing chemical protective suits. While not inexpensive, these efforts would have been critical to keeping NATO's airpower in the fight had war erupted along the old inner-German border.

We strongly suggest that the ROCAF consider adopting this approach to maintaining air base operability. In the near term, the best available answer to the PRC's improving missile forces may be to pour lots of concrete on and around their likely targets, which almost certainly include Taiwan's air bases.

Advanced Air Weapons

Currently, the Chinese and Taiwanese air forces have rough parity in terms of the air-to-air weapons they deploy. Both sides field short-range infrared (IR)-guided AAMs, such as the U.S. AIM-9 or the Russian R-73/AA-11 and medium-range semiactive radar-homing (SARH) weapons, such as the AIM-7 and AA-10. In addition to importing foreign designs, both the PRC and the ROC manufacture IR and SARH AAMs.

SARH missiles require that the launching aircraft illuminate the target during the weapon's fly-out time. This limits the pilot's ability to either engage a second target or engage in offensive or defensive maneuvering and is a distinct disadvantage in air-to-air combat. Currently neither air force has more modern, "fire-and-forget" BVR missiles, such as the U.S. AMRAAM or Russian R-77/AA-12, in its inventory. Both sides are actively pursuing such a capability, how-

Issues and Implications 35

ever, with the Chinese frequently reported as pressing Russia to sell the AA-12.[4]

Figure 3.2 shows the effects of adding BVR capabilities to one or both sides.[5] It shows that the best outcomes for Taiwan occur, as expected, when only the ROCAF fields AMRAAM-type weapons. The first and third bars show that the ROC does fairly well as long as "BVR

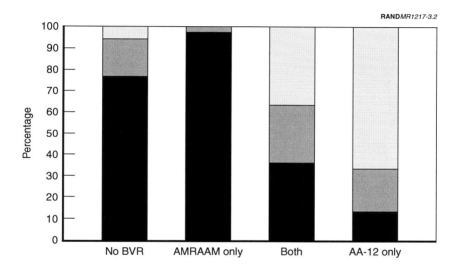

Figure 3.2—Effects of BVR Capabilities on Case Outcome

---

[4]Several reports suggest that Russia has in fact agreed to sell the AA-12 to China (see, for example, Sae-Liu, 1999). Nonetheless, we have yet to see any credible evidence supporting this.

[5]We allowed ROCAF F-16s to shoot AMRAAMs and PLAAF Su-27s, Su-30s, and J-10s to carry AA-12s. In the base case, this gave the PRC a starting total of 96 AA-12 shooters versus 162 AMRAAM carriers for Taiwan. We gave each side sufficient missiles to fight for four days. If one or the other had some missiles but not enough to last for the full four days, we expect that the results would vary more or less linearly toward one of the four depicted cases, depending on how quickly each side ran out. That is, if Taiwan had enough AMRAAMs for two days and China enough AA-12s for three, we would anticipate seeing two days of combat that looked like the first two days of the "both" case, one day that resembled the "AA-12 only" outcomes, and a fourth that broadly resembled the "No BVR" case.

parity" is maintained. Although the proportions change to its disfavor, Taiwan still "wins" about 65 percent of the cases when both sides are equipped with BVR weapons.

The most troublesome case is obviously one in which the PLAAF acquires a BVR capability while the ROCAF does not. Under these conditions, 65 percent of the outcomes were coded as red while less than 15 percent are green. This clearly indicates the importance to Taiwan of maintaining at least "AMRAAM parity"; the recent decision by the U.S. government to provide AMRAAM to the ROCAF if China acquires the AA-12/R-77 is an important and welcome hedge against Taiwan falling behind in this corner of the cross-strait arms race. It also prudently avoids accelerating the introduction of these weapons into the China-Taiwan balance; any unilateral advantage enjoyed by Taiwan would almost certainly be short-lived, and our work suggests that the mainland does better when both have BVR weapons than when neither do.[6] Maintaining the current status quo in this area, then, may be to Taiwan's advantage.

Training Quality

For this study we assumed that Taiwanese aircrew were significantly more capable than their mainland counterparts. We based this judgment on two primary factors:

- ROCAF pilots get significantly more flying time each year than do PLAAF airmen. The training standard for Taiwanese fighter pilots is between 150 and 180 hours per year, while PLAAF pilots may as little as 80 hours in the air each year.[7] That is hardly

---

[6]Even a mediocre pilot can become a serious threat if equipped with an AMRAAM-type missile. Because these weapons are easier to use than older SARH missiles and allow multiple simultaneous engagements, they may reduce the advantage enjoyed by a better-trained pilot in air-to-air combat. A certain minimum competence is still required to employ the missiles effectively, and the superior pilot will continue to have an edge if the fight goes beyond the BVR encounter and becomes a close-in maneuvering battle. However, our finding that China's air force fares better if both sides have advanced AAMs, versus the current situation, is consistent with this hypothesis.

[7]Published sources such as *The Military Balance* credit the ROCAF with 180 hours per year and the PLAAF with 80–110 depending on the type of aircraft. ROCAF personnel we talked with told us that their training levels were around 150 hours per year, while the PLAAF's are often closer to 40–60 hours.

enough flying to remain marginally competent in basic airmanship and navigation, let alone to maintain tactical proficiency.

- While the complexity of PLAAF training and exercises has been improving, the quality of training is also higher for ROCAF aircrew.[8]

That said, we believe that the Taiwanese would benefit from further improving their training. Figure 3.3 shows the results of a set of excursion runs we did (involving no direct U.S. combat involvement) to explore the impact of improved ROCAF training on combat outcomes.[9] The left-hand bar shows that Taiwan achieves a green or yellow result about 70 percent of the time when we trained the ROCAF's aircrew at the baseline 80 percent level. We then added 24

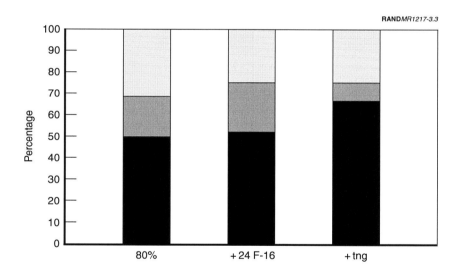

Figure 3.3—Effects of ROC Training on Case Outcome

___

[8]Allen, Krumel, and Pollack (1995, p. 183) report that the PLAAF remains "severely deficient" in many skills, such as low-level flying and dissimilar air combat tactics, that are critical to combat power.

[9]This excursion involved 192 model runs in addition to the 1,700-plus already described.

additional F-16s to the ROCAF force structure; with pilots trained to 80 percent, this produced the middle results, which boosted the proportion of defensive-friendly outcomes to 75 percent. The third bar depicts the results when we trained the ROCAF aircrew to 100 percent (e.g., made them the equivalent of U.S. pilots) but took away the extra F-16s. While the overall percentage of good outcomes increased only slightly over the cases with the added fighters, the proportion of green, or "very good," ones is about 15 percent higher.

The Value of U.S. Involvement

Across the whole range of cases and threats we explored, U.S. involvement made a substantial difference, significantly improving the chances for a successful conclusion to the air war.

Figure 3.4 shows how six alternative packages of U.S. forces affected outcomes.[10] It shows that Taiwanese forces alone had good prospects for a successful defense only about half the time. A single carrier air wing (CVW), or 72 F-15Cs flying from Okinawa, jacks that proportion up to about 65 percent.[11] Combining the two or adding an extra CVW in the absence of the Okinawa-based fighters generated a successful defense in 80 percent of the cases. The biggest U.S. force we examined, two carriers plus the fighters on Okinawa, produced a Taiwan-favorable outcome in more than 90 percent of the cases.

It is important to note that even the largest levels of U.S. commitment under scrutiny here are quite modest compared with those envisioned for a canonical MTW. The United States has twice

---

[10] This discussion assumes that U.S. and Taiwanese forces will be able to at least keep out of each other's way in case of war. This is by no means obviously true today, and we will discuss this in the next chapter when we lay out our recommendations for U.S. action.

[11] It is interesting that we found the impact of either a CVW or the wing at Kadena to be more-or-less the same across all of our cases. We suspect that this stems from the larger number of F-15s (72 versus 48 F/A-18s in each CVW) being offset by the significantly greater distance the land-based fighters had to fly to get to the combat zone. Our analysis did not include Chinese attacks on either the base on Okinawa or the U.S. carrier(s). Such attacks, if effective, could reduce the impact of U.S. forces from that shown here.

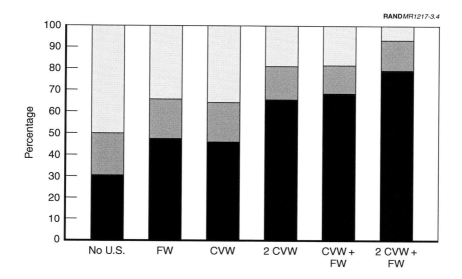

Figure 3.4—Overall Effect of U.S. Involvement

recently (in 1996 and 1999) deployed a pair of carrier groups to the vicinity of Taiwan during periods of heightened cross-strait tension, and the USAF currently bases a full wing of 72 F-15s at Kadena. Yet, these small—by U.S. standards—increments of force appear to make a major difference in the scenario we are considering.

This can perhaps be seen even more clearly in Figure 3.5, which shows the impact U.S. forces have against the most dangerous PRC threat, the advanced case. Almost 70 percent of the outcomes are red if the United States is not involved. A single carrier or the Kadena wing reduces the proportion of bad cases to about half. Combining the two or engaging two carriers by themselves yields defense-favorable results about 55 percent of the time and provides an even more dramatic increase in the number of green cases. Finally, our largest U.S. force commitment (two CVWs and one land-based FW) helps provide Taiwan with a positive outcome in almost 80 percent of the cases we studied against the advanced PRC threat.

Although the impact of U.S. forces is most dramatic when Taiwan confronts the most severe threats, they can play a significant role

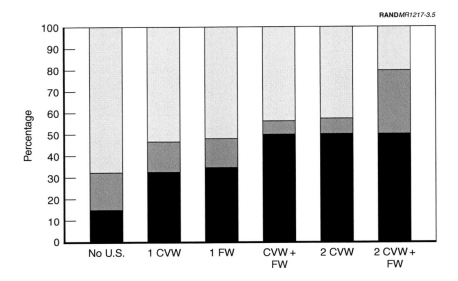

Figure 3.5—Effect of U.S. Involvement on Air Outcomes, Advanced Threat

even against the base PLAAF force. Taiwan on its own achieved good outcomes against the base PRC threat in about 65 percent of our simulations but "won" almost 95 percent of them—a 45 percent improvement—when U.S. forces were also engaged.

In sum, then, U.S. forces can make a considerable contribution to Taiwan's ability to maintain control of the air against all three PRC threats. The United States role appears to become increasingly critical as the PLAAF commits more assets to the operation or becomes more sophisticated.

MARITIME SUPERIORITY

The ASW Dilemma

The Taiwan Strait is a notoriously poor environment for ASW operations. The waters are shallow, heavily trafficked, and generally provide unreliable acoustic propagation. This works heavily in the PLAN's favor, because the mainland's single biggest naval advantage over Taiwan lies in its large fleet of submarines.

It is important not to overestimate the potential impact of China's subs. Most of its underwater order of battle consists of vessels of dubious quality—Chinese copies of obsolete Soviet designs. Further, the training level of the PLAN's submariners, and the operability level of their boats, is widely regarded as subpar. Nonetheless, China's enormous numerical advantage coupled with the difficulty Taiwan would have in prosecuting an ASW campaign in the strait could figure heavily in the battle for maritime control. In many of our model runs, Chinese submarines sank a considerable portion of the ROC surface fleet in the opening minutes and hours of the war.

Our analysis found no simple solutions to this dilemma. Taiwan could try to rely on fixed- and rotary-wing aircraft to level the underwater playing field somewhat. However, if the intensity and ferocity of the air war were anything like what was seen in our simulations, it seems unlikely that ASW aircraft would be able to operate effectively. The ROCN's more modern surface combatants, such as the *Cheng Kung* and *Kang Ting* frigates, have fairly up-to-date ASW suites, but China's *Kilo* submarines, if properly operated, could prove elusive prey in the difficult environment of the strait.

Faced with this problem, Taiwan might want to consider a naval strategy analogous to the "rope-a-dope" approach to boxing made famous by legendary heavyweight Muhammad Ali. Confronting a stronger but slower adversary, Ali would linger near the edges of the ring, avoiding many punches and deflecting the others and waiting for his opponent to tire. Similarly, we found that the ROCN achieved its best results when it sought to overextend and exhaust the adversary by keeping the bulk of its surface forces out of the strait during the initial hours and days of a confrontation with the PRC. If the Chinese were to attempt to push their submarines further out to engage the ROCN, they quickly found themselves beyond the PLAAF's air umbrella, exposed to attack by Taiwanese ASW aircraft and helicopters. And, the performance of both airborne and surface ASW systems would improve significantly outside the confined and difficult waters of the strait.

If, on the other hand, the PLAN declined to go after the ROCN, it would face a reasonably intact "fleet-in-being" when it came time to actually move any invasion force across from the mainland. The Taiwanese would still have to cope with the Chinese submarine threat when the ROCN "rushed" the strait. However, compared with a situation where the two navies have sailed eyeball-to-eyeball in the strait while tensions turned to hostility, it seems much less likely that the Chinese would be able to place their subs in ideal attack positions to target a Taiwanese "rush" against the invasion fleet. The ROCN would likely suffer heavy losses even employing the "rope-a-dope" strategy, but enough combat power could survive to significantly disrupt the Chinese assault.[12]

Maintaining a Credible Antisurface Warfare Capability

If, however, the main battle forces of the ROCN stay out of the strait during the initial phase of the war, what sort of resistance could the ROC put up against the PLAN's opening moves? Here, we suggest that Taiwan can capitalize effectively on the inherent strength of its defensive position.

The ROCN is currently in the early stages of procuring up to 50 *Kuang Hua VI* missile boats. This vessel is planned to be small, stealthy, and well-armed with four antiship missiles. Operating close to Taiwan's coastline, where search and targeting radars will have difficulty discerning targets amid the clutter, the *Kwang Hua VI* class could prove a potent source of firepower even if the ROCN's frigates and destroyers were not in the vicinity. Indeed, the absence of large ROCN warships could reduce problems with identification, friend or foe (IFF), and improve Taiwan's overall ability to locate, target, and engage PLAN surface ships.

Stealthy, lethal missile boats could be made even more effective in combination with shore-based defenses. Mobile antiship missile launchers, well-protected against Chinese SOF, would provide additional firepower to Taiwan's initial defense. These missiles should,

---

[12]Taiwan's difficulties in the ASW area may also impact the kinds of solutions that are appropriate for its future ballistic missile defense (BMD) system. Any sea-based platform that needed to be in the strait to be effective would become an obvious target for China's submarines and would be at great risk for the reasons outlined above. Conversely, a system that could retain its punch while standing off north, south, or east of the island would gain significantly in survivability.

for the near future at least, greatly outrange the PLAN's guns, allowing them to be highly survivable in a "shoot-and-scoot" mode.[13]

Targeting will potentially be a problem for both missile boats and shore defenses. Fixed radar sites will almost certainly be prime targets for Chinese ballistic missiles in the first wave of attacks. Mobile surveillance systems, including the ROCAF's E-2T Hawkeye radar aircraft, may be more survivable but could be at least partially neutralized by Chinese jamming. If the radars can operate, high-speed sensor-to-shooter data links will be needed to ensure that fires are delivered in a timely manner. Keeping Taiwan's navy largely off the battlefield will simplify these problems—any big ship "out there" will be Chinese—but the ROC must consider how best to build a robust and highly responsive C^2 intelligence, surveillance, and reconnaissance network to support whatever antisurface warfare strategy it proposes to follow.[14]

The U.S. Role

We have already touched on the contribution that one or two U.S. carrier battle groups could make to the air defense of Taiwan. In the time frame we are considering, the Chinese will also have no reliable counter to the maritime warfare capabilities of a CVBG. In several war games conducted using *Harpoon*, two U.S. CVBGs operating east of Taiwan were able to decimate the PRC's surface and submarine fleets even when the PLAN's best combatants—*Sovremenny* destroyers and *Kilo* submarines—were put in advantageous firing positions and allowed to fire the first salvo at H hour.

In our assessment of the U.S role in the maritime campaign, three functions in particular caught our imaginations.

---

[13]Similarly, advanced mobile air defenses like man-portable missiles and the Avenger system currently fielded by the ROC Army could be very valuable counters to Chinese airborne or heliborne assault troops. Taiwan should distribute these weapons widely throughout the ground forces.

[14]At the risk of being redundant, we also wish to note that neither missile boats nor shore-based missile launchers, nor even the most mobile of surveillance radars are likely to last long if the PLA has achieved any significant degree of air superiority. Thus is the outcome, or progress, of the air war a determining factor in the sea battle.

First, any submariner worth his salt will say that the best ASW weapon is another attack boat.[15] The SSNs that travel with most U.S. CVBGs completely outclass anything that the PLAN is likely to field in the next 20 years, let alone five. One or two modern U.S. submarines operating in the Taiwan Strait could redress Taiwan's shortfalls in ASW capability, provide vital intelligence support to the ROC, and put formidable antisurface warfare striking power just off the Chinese coast—if appropriate rules of engagement could be worked out.

For example, the United States and Taiwan could "quarter" the strait, with one ROCN sub in each of the sectors near the Taiwanese coast and two U.S. SSNs divided between two sectors nearer the mainland. Each sub could then treat its zone as a weapons-free zone for ASW hunter-killer operations. Requirements for coordination between ROCN surface forces and U.S. submarines could be minimized by simply keeping ROCN combatants east of the center line of the strait, if indeed they are in the strait at all.[16]

Second, the United States could also help provide surveillance and targeting for the surface battle. USAF E-3 AWACS aircraft operating from Guam or Japan, along with Navy E-2 Hawkeyes, could supplement Taiwan's radar picture of the battlespace, while other intelligence collectors—such as electronic and signals intelligence (ELINT and SIGINT) platforms—could likewise augment the ROC's capabilities in these areas. Again, the ability to rapidly establish and maintain the appropriate data links would be all-important in ensuring that the right information flowed to the right places in time to be exploited.

---

[15] Historically, submarine-versus-submarine combat has been the exception, not the rule. However, in past wars that featured extensive submarine operations, subs were essentially surface ships that could submerge for brief periods to prosecute an attack or to escape one. Submarine deck guns were as important as torpedoes, and subs were most commonly lost to air attacks on surfaced or diving vessels. Modern submarines, whether conventional or nuclear-powered, operate almost exclusively submerged, which dramatically changes the nature of ASW. Virtually all advanced navies, including the U.S. Navy, rely on their own submarines as a primary weapon against an enemy's undersea forces.

[16] Presumably, Taiwan's surface and subsurface naval forces would have many prewar opportunities to work out procedures to minimize the chances of accidentally engaging one another.

Finally, the USAF maintains a number of B-52H bombers that are equipped and trained to deliver Harpoon antiship missiles. With a range of more than 70 nautical miles, the sea-skimming Harpoon could be fired from outside the range of most Chinese surface-to-air defenses (and all of its ship-based SAMs) and would likely prove lethal against the weak AAW systems fitted to PLAN combatants (not to mention the nonexistent air defenses of the commercial shipping that could constitute the bulk of any invasion fleet). Again, keeping large ROCN surface warships out of the strait would prove advantageous in reducing the likelihood of fratricide from such strikes.

SUMMING UP

It may be useful at this juncture to recap the main points.

In terms of the pivotal fight for air superiority, four factors emerge as critical in our analysis:

- *Taiwan's air bases must remain operable so that the ROCAF's fighter force can keep up the battle against the PLAAF's superior numbers.* We recommend increased attention to passive defense and rapid-reconstitution measures.

- *The ROC must maintain at least parity in advanced air-to-air weaponry.* Ideally, Taiwan could enjoy a unilateral advantage in this area; realistically, however, the present status quo may represent the best situation for Taiwan. The U.S. decision to provide AMRAAM if the PRC receives AA-12s is an important and useful hedge.

- *Pilot quality may be Taiwan's ace in the hole.* PLAAF training is notoriously poor. This is even more reason for Taiwan to ensure that its aircrews are of the highest possible caliber. Our analysis suggests that improved pilot quality may contribute more to favorable air superiority outcomes than would even sizable additions to the ROCAF's fighter force structure.

- *U.S. involvement is important now and will likely grow increasingly vital.* Even in the near term, U.S. carrier- and land-based fighters could make a combination crucial to Taiwan's defense. As the PLAAF's inventory becomes more sophisticated and capable, Taiwan's need for U.S. assistance will likewise increase.

We can cite three main insights into the naval war in and around the Taiwan Strait:

- *ASW is a critical Taiwanese weakness.* Absent an unexpected acquisition of numerous modern attack submarines, the ROCN will have tremendous difficulty coping with China's modernizing submarine fleet. We suggest that the ROCN consider keeping its main battle forces out of the strait during the initial phase of a war with the mainland.

- *Fast, stealthy missile boats and mobile, land-based antiship missile launchers can help Taiwan exploit its inherent defensive advantages.* If adequate detection and targeting information can be provided, these weapons could prove highly lethal and relatively survivable even in the chaotic opening hours of a China-Taiwan clash.

- *Again, the U.S. role in the naval campaign could be crucial.* U.S. SSNs could help counter the Chinese submarine threat, U.S. surveillance capabilities could provide vital support to Taiwanese forces, and Harpoon-equipped bombers could provide key early firepower to the naval battle.[17]

The next and final chapter of this report attempts to tie our findings together into recommendations for near-term U.S. policy on supporting Taiwanese security.

---

[17]Navy F/A-18 strike aircraft can of course also carry Harpoons. However, in the early stages of a China-Taiwan war, we suspect that these aircraft would find their best use in the air-to-air battle. Therefore, we focus on the possible role of long-range, land-based bombers for delivering antiship strikes in the first few days.

Chapter Four

RECOMMENDATIONS AND CONCLUDING REMARKS[1]

While our work has naturally resulted in several recommendations of ways that Taiwan could enhance its self-defense, our stated purpose is to inform U.S. policy with regard to supporting the ROC's security needs, and it is on this topic that this chapter is focused.

U.S. SUPPORT IS VITAL TO TAIWAN'S SECURITY[2]

Ideally, of course, the conflict we have studied will never occur. Given, however, that it seems unlikely that Beijing will renounce its "right" to use force to achieve unification, a strong Taiwanese deterrent appears to be an important component of continued peace on the strait. As Taiwan's most reliable friend and in keeping with the requirements of the 1979 Taiwan Relations Act, the United States will

---

[1] The reader will no doubt notice that this chapter makes recommendations based primarily on operational military grounds with only passing reference made to the difficult political context of U.S.-Taiwan security relations. This is in keeping with our purpose, which is to derive insights regarding Taiwan's security from an analysis of the military aspects of the confrontation between the PRC and the ROC. Our relative neglect of the diplomatic and political aspects of the situation should not be interpreted as a disparagement thereof. We recognize their centrality and importance. Our contribution to the debate simply lies elsewhere.

[2] We recognize that there are many constraints on U.S. support for Taiwanese self-defense. Most obvious are the political sensitivities that surround all aspects of U.S.-Taiwan relations. However, operational issues exist as well, with U.S. military leaders occasionally expressing reluctance to provide Taiwan with weapons that could inadvertently threaten U.S. forces should they be called on to come to Taiwan's defense in some future contingency. We are indebted to Michael Swaine for bringing this to our attention.

necessarily play a major role in helping the ROC maintain and enhance its defensive capabilities even as the PLA modernizes.

Keeping the peace on the Taiwan Strait is important for the United States. A war between China and Taiwan, regardless of the outcome, would be a disaster for both sides, for East Asia as a whole, and for the United States. At the same time, care should be taken that steps to increase Taiwan's defensive capability not be misread by Beijing; neither Taipei nor Washington has anything to gain by provoking the PRC to precipitate actions.

Should deterrence fail, Taiwan may in some circumstances find itself in a position where its survival depends on some degree of direct U.S. military intervention. We do not wish to prejudge whether or not such an action would, or should, be forthcoming. Our analysis, however, suggests five key insights regarding U.S. support for Taiwan—in both peace and war—that indicate ways of enhancing deterrence across the strait. By pursuing initiatives along these lines, Taiwan's defense posture vis à vis China could be significantly enhanced with, we believe, minimal risk of destabilizing the situation.

Small Increments of U.S. Assistance Could Turn the Tide

First, the amount of force needed to support Taiwan in the near term appears to fall considerably short of what is usually thought of in the Pentagon as that which is needed to prosecute an MTW. In the results we report here, we never committed more than a single wing of land-based fighters, two CVBGs, and a dozen or so heavy bombers to the campaign, a far smaller force than the 10 fighter wing equivalents and six CVBGs that engaged in Desert Storm.[3] The sheer compactness of the combat zone around Taiwan limits the number of forces that can effectively be employed, and we found that even a very small U.S. contribution—one or two carriers, a couple of squadrons of F-15s, and two SSNs—could frequently swing the outcome in Taiwan's favor. The balance between the mainland and Taiwan is fairly fine and should remain so for the next five years or so; *U.S. thinking about a PRC-ROC scenario should focus on rapidly*

---

[3]Never mind the two Army corps and two Marine divisions that fought the ground war in the desert.

bringing limited force to bear rather than a full-up mobilization of the vast American military machine.[4]

Supporting Taiwanese Modernization: The Israel Model

For many years, the United States has been the primary external supporter of Israel, an actor that, like Taiwan, has lived under constant threat of attack by much larger neighbors. In terms of arms sales and military assistance, U.S. policy has revolved around the idea that Israel should maintain a qualitative superiority of sufficient degree to offset its numerical inferiority in a defensive campaign. We suggest that a similar philosophy could guide U.S. assistance to Taiwan. How best to implement such a strategy is an important question.

Taiwan currently fields a number of first-rate weapons platforms. The F-16C and Mirage 2000, for example, are at least the equals of any fighter in the PLAAF inventory, and the ROCN's modern frigates are likewise a match for the mainland's best surface warships. Helping Taiwan exploit these systems to their fullest might be the best way for the United States to enhance the island's defensive capabilities.

There appear to be two primary paths to such improvement. The first would involve U.S. assistance in acquiring or developing armaments, sensors, and other equipment that will help maximize the capabilities of Taiwan's platforms. We have already stressed, for example, the criticality of "AMRAAM parity" between the PRC and the ROC. Following the recent U.S. decision to sell but not deliver AMRAAMs to Taiwan, the necessary political and technical groundwork should be laid to permit the ROCAF to quickly assimilate the weapon if and when the PLAAF begins to field the AA-12. The United States could also ensure that the Taiwanese military has access to advanced aerial surveillance radars as well as modern air defense C^2 systems. Upgrades to the ROCAF's E-2T fleet of radar aircraft to make them more capable and more reliable could be especially valu-

---

[4]If the initial defense of Taiwan should fail, a much larger U.S. commitment would almost certainly be needed if the decision were made to "liberate" the island. Such a scenario is far outside the ken of this study.

able, because these platforms may be more survivable than fixed land-based radar sites. Finally, the United States could help the ROCAF protect its air bases against the kinds of attacks that the PRC could mount against them. As we noted above, sustaining air operations under attack—or recovering rapidly after one—would be vital to the ROCAF in any clash with China. The United States could assist by providing equipment, material, and know-how.

At sea, Taiwan could benefit first and foremost from enhancements in ASW. Advanced towed-array sonars, signal processing equipment, modern ASW aircraft, and improved weaponry could all prove valuable in helping the ROCN deal with the PLAN submarine threat.[5]

The second way the United States can help Taiwan get the most out of its inventory of advanced weapons is to assist in efforts to improve the quality of the island's military personnel. This could include assisting the ROCAF in acquiring a larger number of advanced flight simulators, for example, or helping the Taiwanese develop improved training programs. Such assistance need not be limited to flying personnel, either; improving the competence of ground crews and maintenance technicians will help assure that the ROCAF can maintain the high sortie rate needed to combat China's far superior numbers.

Direct training contacts between U.S. and Taiwanese military personnel always carry the risk of raising Beijing's ire. Nonetheless, we

---

[5]Taiwan has expressed great interest in acquiring some number of AEGIS-equipped warships similar to the U.S. Navy's *Arleigh Burke* class of destroyers, and the enhanced air defense capabilities of these ships would certainly be valuable. However, these ships, if imprudently employed, would be very lucrative targets for Chinese submarines; their ASW capabilities, after all, were intended to be but one component of the overall defenses of a U.S. carrier battle group. As an alternative, we suggest that the United States and Taiwan focus on incrementally improving the ROCN's existing ship-based air defenses. Actions worth considering could include fielding better sensors, improving the networking of land- and sea-based systems, and deploying more-capable missiles. The launchers on Taiwan's *Cheng Kung* frigates, for example, can, with modest modifications, accommodate the longer-range and more-capable Standard SM-2 round as well as the SM-1 variant with which the vessels are currently equipped. As the SM-1 is out of production, switching to the SM-2—which may involve refitting fire-control radars as well—might become necessary if the ROCN is not to lose its extended-range AAW capabilities. In the mid- to long-term, the need to replace Taiwan's aging destroyers may require the acquisition of warships in the *Burke* class.

believe that some expansion of these activities would have great value. Allowing even a few ROCAF pilots to participate in advanced USAF training events, such as Red Flag, or attend the Fighter Weapons School, could in the mid- to long-term have a major impact on the ROCAF as the experience and knowledge they gained percolates through the force. Even less visible, and hence perhaps more palatable, increases in interaction—such as permitting more Taiwanese officers to attend such U.S. professional military schools as the Air War College—could pay off.

Air Defense C^2

While sales of platforms and major weapons systems grab the headlines and generate the controversy, Taiwan has critical needs in other areas. The air defense C^2 network, which has been upgraded substantially in the past decade, continues to suffer from limitations in intelligence fusion and data transmission and relies on vulnerable fixed radars for much of its situational awareness. The E-2T radar aircraft offer a mobile, potentially survivable adjunct to ground sites but lack the kinds of data links needed to exploit their capabilities fully in an integrated fashion. These shortcomings should be an important priority for rectification. The U.S. military and the American private sector can offer Taiwan the world's finest systems engineering and integration solutions in terms of both hardware and software. The U.S. side can encourage Taiwan to make the investments needed to ensure that the ROC's C^2 system is fully modernized and robust in the face of the kinds of threats it would likely face in a conflict with China.

Information and Intelligence Sharing

The United States is obviously and properly sensitive and selective in choosing how and when to share what kinds of information and intelligence with its friends and allies. At the same time, however, there would appear to be enormous leverage to be gained by helping Taiwan's government and military leadership maintain an accurate picture of the strategic and tactical situation day to day and, especially, during a crisis.

During peacetime, Taiwan requires an up-to-date and unprejudiced perspective on China's political and military planning, intentions, and capabilities. Its own intelligence service is naturally the primary source for the ROC's insights into Beijing's likely actions and the state of the PLA, but cooperation with the United States in this field could improve the amount and quality of information available to decisionmakers in both Taipei and Washington. A shared picture of the evolving threat would also likely make it easier for the two sides to reach agreement on arms sales and other modes of U.S.-Taiwan defense cooperation.

In a crisis, U.S. intelligence support could be critical to Taiwan's security. U.S. collection systems could provide early warning of force movements, such as the marshaling of airborne troops and transports, deep inside China that could presage a move against Taiwan. Conversely, a failure to detect such necessary preparatory actions could help defuse a crisis should Beijing attempt to intimidate Taiwan by bluffing with highly visible maneuvers. In any event, political and military leaders on both sides of the U.S.-Taiwan relationship stand to gain a clearer common image of any emerging situation if the two sides routinely and regularly share intelligence and information.

Interoperability: The Critical Link

Here we come to a crucial task: ensuring that U.S. and Taiwanese forces can fight side by side, should the need ever arise, without introducing crippling levels of friction into either's operations. This area is of very serious concern.

In this analysis, we assumed that the United States and Taiwan had achieved only a minimum level of interoperability—that the airspace over Taiwan and the strait could be divided and U.S. and ROC forces segregated into their own operational sectors. We further assumed that the U.S. Navy and USAF were unable to directly link their surveillance and targeting systems, such as AWACS and AEGIS, to their Taiwanese counterparts and vice versa. Essentially, we tried to keep the ROC military and the U.S. military out of each other's way while each fought more-or-less independent campaigns against the Chinese.

This may seem to be a pessimistic picture, but it may in fact overstate the degree of cooperation that would be possible if war were to break out today. Last year's operations over Kosovo and Serbia demonstrated the very real and binding limitations of interoperability that exist even within the 50-year-old NATO alliance. Despite decades of the closest possible interaction and joint training and despite volumes of standard operating procedures and standing agreements, alliance air operations were impeded by technical incompatibilities, procedural differences, and even language difficulties. Absent this corpus of common experience and planning, how effectively could the U.S. military actually expect to fight with the Taiwanese?

There are both hardware and software aspects to this problem. Many Taiwanese weapons—even those provided by the United States—lack specific capabilities, such as data links and IFF systems, that would enable them to "plug and play" with comparable U.S. platforms. Even when physical equipment is identical, computer codes have sometimes been embargoed, resulting in incompatibilities. Some such modifications are justified: Taiwanese F-16s certainly do not require the wiring needed to mount and deliver nuclear bombs, for instance. However, decisions regarding what to leave in and what to leave out could take into account the possibility that the F-16s (or radars or SAMs or shipboard combat-control systems) in question could someday be employed in conjunction with U.S. forces. Prudence seems to pull in both directions on this issue, but we believe that enhancing the ease of cooperation between Taiwanese and U.S. forces is in the interests of both sides.[6]

Small and discreet steps could be valuable. Developing the basic framework for joint airspace management, for example, need not involve high-profile, high-level visits. Upgrading software to improve compatibility could be done almost invisibly. Indeed, to the extent that operational capabilities are coming to reside less and less in the basic platforms and instead stem from the bits and bytes

---

[6]Beijing would find visible steps in this direction extremely annoying, which is regrettable and clearly a factor to be considered in determining precisely what to do and how far to go. However, it is also testimony to the potential deterrent value of improving interoperability: if it didn't make a difference, the Chinese likely would make much less of a fuss.

flowing through "black boxes," it may become less politically painful to engage in this sort of cooperation.

The human side of the equation should not be neglected, either. Although it seems unlikely that the United States and Taiwan will be engaging in large-scale bilateral exercises over the next five years, contacts between U.S. and Taiwanese planners and operators could increase. Working-level discussions between the men and women who would actually be put on the spot in any conflict would be invaluable in paving the way to smoother interactions in the event.

This is obviously an extremely sensitive area and one where effects are prodigiously difficult to quantify. Nevertheless, based on our discussions with both the U.S. and Taiwanese militaries, we are convinced it is a problem that needs to be addressed, and even small steps could make a significant difference.

CHINA AS A SANCTUARY?

If an all-out war between Taiwan and China should erupt, the ROC government will face a dilemma regarding whether or not to strike targets on the Chinese mainland. Taiwan has shown at least some interest in developing its own surface-to-surface missiles; however, no confirmation exists that it has deployed any such systems (Daly, 1999, pp. 24–29). We do not wish to speculate whether or not Taipei would choose to take offensive action or to argue one way or the other about its advisability.

It is important to note, though, that the United States would likely face the same conundrum if it were to become actively involved in defending Taiwan. As demonstrated in Iraq and again in the Balkans, contemporary U.S. warfighting strategy typically includes large-scale strikes against command, control, and communications facilities, air defenses, air bases, and an array of other targets in the adversary's territory. Whether or not the United States would initiate such a campaign against a vast and nuclear-armed opponent—and, if so, what sorts of limitations would be imposed on targeting and collateral damage—is a deeply vexing question.

The need to suppress the PLA's long-range air defenses could provide the most compelling rationale for at least limited attacks on military targets in China. Beijing currently deploys a small number—a half-dozen or so batteries—of Russian S-300PMU1/SA-10D SAMs, whose 48N6 missiles have a maximum range estimated at 150 kilometers. Further developments of this family of missiles have been reported with ranges extending as far as 400 kilometers; China itself is developing its own spinoff of the basic SA-10, the HQ-9. Should the PLA locate several of these systems close to the Chinese coast, they could cover much of the strait at medium altitudes and even reach over Taiwan itself at high altitudes. Low-level tactics would reduce the S-300's engagement range but would also restrict the tactical and operational flexibility of Taiwanese and U.S. aircraft. Neutralizing these long-range "double-digit" SAMs is widely regarded as a difficult tactical problem. Adding in the risks associated with attacking even strictly military targets within China compounds the complexity.

We have no answer on this; we suggest, however, that U.S. planners think through in advance the implications for U.S. operations of either striking or not striking military targets in China. The stakes are too high, both on the battlefield and at the strategic level, to leave considering the question to the last minute.

LOOKING BEYOND 2005

This study was exclusively focused on the near term and included only capabilities that could conceivably be fielded by 2005.[7] Naturally, however, the analysis leads to some points about possible evolutions beyond this time frame and their potential implications.

A number of developments on the Chinese side emerge from our assessment as appearing particularly troublesome. These include the following:

- Advances in information warfare capabilities that enable China to more rapidly and completely shut down Taiwan's C^2 networks.

---

[7]Although, as noted earlier, we may have given the Chinese in particular the benefit of several doubts in our efforts to be conservative.

- The deployment of hundreds or thousands of conventionally armed and highly accurate ballistic and cruise missiles that could greatly endanger the operability of Taiwan's air bases.
- Fielding of a standoff munition similar to the U.S. JSOW that would enable the PLAAF to deliver ordnance accurately onto many Taiwanese targets from within or just outside the coverage umbrella provided by China's long-range SAMs.
- Large numbers of GPS-guided free-fall munitions (akin to the U.S. JDAM) that might turn older aircraft with poorly trained pilots into reasonably effective attack platforms.

Countering such developments would present new challenges to the Taiwanese. Looking toward this uncertain future, we recommend that the United States work to help Taipei improve its ability to defend key military and commercial information systems from attack. Also, with the Chinese likely to exploit GPS and Russian GLONASS navigation satellites in the guidance modes for many future weapons, Taiwan may want to acquire the ability to effectively jam these signals over both its own territory and the strait. Both the ROC military and the United States may therefore want to invest time in planning both how best to operate in such a "GPS-out" environment and fielding highly jam-resistant equipment to sustain GPS use in a very difficult electronic environment.

FINAL THOUGHTS

A war between Taiwan and China is an unpleasant and tragic prospect, rife with the potential for escalation and guaranteed to result in massive destruction both in the immediate military sense and in longer-term damage to the political evolution of East Asia and the Pacific Rim.

This study suggests that Beijing would be imprudent to resort to massive air and missile attacks or an invasion of Taiwan as a means of compelling unification. Our results show an incredibly costly war that the PLA should have serious doubts about winning. The odds against the mainland appear to increase still further if the United States gets actively involved—even minimally—in Taiwan's defense.

Maintaining peace on the Taiwan Strait is in the interests of the Chinese people on both sides of the narrow waters and of the American people, too. To the extent that China continues to threaten military action against Taiwan, deterrence will remain an important component of any strategy aimed at avoiding conflict. Sustaining and enhancing that deterrent—which boils down to sustaining and enhancing Taiwan's defensive capabilities—is a crucial goal of U.S.-ROC security cooperation.

That this objective is purely defensive should be emphasized publicly and be clearly visible in the types of weapons sold, technology transferred, and assistance lent. While the dichotomy between "offensive" and "defensive" stances can be easily overdrawn, initiatives that clearly respond to the developing Chinese threat—such as helping Taiwan improve its air base operability, increasing the survivability of its C^2 systems, or providing AMRAAM when the PLAAF begins deploying AA-12s—can be strongly justified and are good candidates for near-term implementation. To do little or nothing in the face of growing Chinese capabilities would diminish the deterrent posture that is critical to keeping the waters between Taiwan and the mainland from becoming a "dire strait" indeed.

Appendix A
SOME THOUGHTS ON THE PRC MISSILE THREAT TO TAIWAN

The PRC's surface-to-surface missile forces would be a major factor in an invasion scenario, such as the one we assess in the main body of this report. China's missiles would also be at the heart of many other coercive strategies aimed at Taiwan, ranging from the kinds of exercises held in 1995 and 1996 to actual employment against military, political, and economic targets in an attempt to use limited force to compel Taiwanese concessions. Indeed, it is widely believed that missile attack represents the greatest strategic threat currently confronted by Taiwan.[1]

There are at least four reasons why Taiwan should be deeply concerned about the missile threat. First, effective defenses are not likely to be available much before 2010. Current and near-term systems, such as the Patriot PAC-2 and PAC-3 and the U.S.-Israeli Arrow, are designed to intercept Scud-like missiles and may have only limited capability against the more advanced weapons entering

---

[1]It is important to keep this threat in context. The largest number of missiles we employed in any of our scenarios was 620. According to open-source estimates of their payloads, their total "throw weight" would be about 450 tons of high explosives. To put this in perspective, the average operational bomb load of a World War II B-17 was about two tons. By 1945, the U.S. Eighth Air Force was routinely dispatching formations of 600 to 1,000 Flying Fortresses to attack individual industrial complexes in Germany. By the yardstick, the 600-plus Chinese missiles we aim at Taiwan could deliver about one-quarter the explosive power of a single "thousand-plane raid" on the Third Reich. While this is a nontrivial amount of destructive power, it is far from clear that it would be in fact sufficient to materially devastate Taiwan's military, economy, or society. The potential psychological impact of such attacks, of course, should not be underestimated.

the Chinese inventory. And, all these systems are dependent on cumbersome surveillance radars that are highly vulnerable to attack (by SOF, for example) and difficult and expensive to replace.

Second, the threats will continue to grow in number and improve in quality. According to one estimate, China has the "industrial capacity" to produce up to 1,000 new missiles over the next eight to 10 years, and some reports suggest that up to 650 of them could be aimed at Taiwan (U.S. Department of Defense, 1997, p. 4; Gertz, 1999, p. A12). A recent Department of Defense report concluded:

> By 2005, the PLA likely will have deployed two types of SRBMs and a first-generation LACM [land-attack cruise missile]. An expanded arsenal of accurate conventional SRBMs and LACMs targeted against critical facilities, such as key airfields and C^2I nodes, will complicate Taiwan's ability to conduct military operations. (U.S. Secretary of Defense, 1999, p. 5.)

Third, international response to limited missile attacks might be less drastic and widespread than that which would be called forth by an outright invasion. China could try to moderate global opprobrium by directing its attacks primarily at military targets, weakening Taiwan's self-defense capabilities in the process.

Finally, the psychological impacts of missile attacks are unpredictable. There are, however, precedents for dramatic impacts. In 1944, Allied operations in Europe were greatly affected by British fears regarding the potential public reaction to German V-2 rocket attacks. Even with the war essentially won, the Churchill government worried that popular support could be threatened if the attacks were not stopped or at least reduced in intensity. So, air and land campaigns in the European theater were redirected so as to more quickly overrun or destroy known V-2 launch areas and production facilities.

More recently, the "war of the cities" between Iran and Iraq marked the terminal phase of their drawn-out and bloody war. Despite what was essentially a stalemate on the ground, Tehran was induced to accept Iraqi terms for a cease-fire at least in part because of the effect that Scud attacks were having on Iran's urban populations. Already worn out by almost a decade of war, many Iranians simply could not bear the persistent terror attacks.

A problem of this magnitude demands a multipronged response. We wish to propose two elements of such a strategy.

First, Taiwan could obviously consider acquiring effective theater missile defense (TMD) systems as they become available. Indeed, even systems that might only be modestly effective, such as PAC-3, could be useful in two ways. By thinning out Chinese strikes, they would reduce the damage absorbed by Taiwan, and—as demonstrated so powerfully in Israel during the Gulf War—they provide a significant psychological boost to the people under attack.

However, TMD is not a panacea. Any defense that relies on unhardened fixed installations, such as radars, risks being disabled by enemy action.[2] Further, all proposed TMD architectures can be overwhelmed by a sufficiently large barrage of missiles. China has in fact threatened to respond to any deployment of TMD in Asia by greatly increasing the size of its short-range missile forces.

As a second element in our strategy, then, we return to a theme we sounded earlier: Taiwan should invest in making important military targets largely invulnerable to missile attacks. This can be done through hardening, mobility, and redundancy. Concrete and construction labor are relatively cheap compared with TMD systems or F-16s. Burying fuel storage at air bases, relying on underground pipelines instead of tanker trucks to deliver fuel to hardened shelters, acquiring mobile air-defense surveillance radars, and creating multiple, mutually supporting data links and C^2 networks are all highly effective antimissile defenses. Also, by making the ROC military passively self-protecting, this approach would allow such active TMD systems as can be fielded to be concentrated around key economic, political, and cultural targets that cannot be buried under reinforced concrete or moved around the countryside.

The psychological effects of a rain of Chinese missiles on Taiwan could be devastating if the Taiwanese people feel unprepared and unprotected. By preparing itself to absorb attacks without large-

---

[2]The same holds true for such large, vulnerable air platforms as the Boeing 747 that will carry the USAF's airborne laser. And, as we pointed out in Chapter Three, ship-based TMD systems would immediately go to the top of the target list for the Chinese Navy (especially the PLAN's submarines).

scale losses of capability, the ROC military can help mitigate this impact. In the near term, the main role for the Taiwanese armed forces in antimissile operations will be to sustain their ability to defend the island against invasion and occupation. Maintaining a robust overall defensive posture will reassure both the Taiwanese population and the world community at large that the ROC retains both the ability and the will to protect itself.[3]

The United States would almost certainly be a critical partner in any Taiwanese effort to develop or procure TMD capabilities. This area is fraught with difficult political questions; the PRC has clearly stated its opposition to Taiwan's inclusion in any U.S.-sponsored missile-defense regime in East Asia. Whether or not Taipei is ultimately successful in fielding active TMD, the United States can assist Taiwan in responding to the missile threat in at least two other ways.

First, the United States could make clear that it would view any use of military force against Taiwan, however limited, as a grave step that would have immediate and serious consequences in Sino-U.S. relations. This is *not* the same thing as making an open commitment to the defense of Taiwan. It is instead simply avoiding the mistake that arguably was made before both the Korean War and the Gulf War: that of signaling to the aggressor that the United States would likely stand aside from any military confrontation. While "strategic ambiguity" may have its advantages as a declaratory policy, no one in Beijing should suffer any illusions that relations with Washington would be unaffected if it unleashed the PLA against Taiwan.

Second, at the risk of being repetitive, the United States can encourage and materially support an immediate and intensive Taiwanese program of passive protection against air and missile attack. This program should focus on military targets, but might also be extended to political and economic facilities that could benefit from cost-effective forms of hardening and increased security.

---

[3] Militarily, Taiwan might be inclined to develop offensive weapons that could enable a "tit-for-tat" response to any PRC attacks. However, the vast difference in size between the two countries suggests that this would be an ineffective response, and the likely political fallout may make it an inadvisable one. Jeopardizing U.S. support to field a token missile force would simply be a bad, bad bargain for Taipei.

Appendix B
OVERVIEW OF THE JICM

The JICM (Joint Integrated Contingency Model) is a game-structured simulation of major regional contingencies, covering strategic mobility, conventional warfare in multiple theaters, and naval warfare. It is a deterministic model with a four-hour time step.

The JICM air war is organized around an ATO (Air Tasking Order), which explicitly packages sorties at the start of each day to execute across the six four-hour periods. The JICM models a number of different air-to-air and air-to-ground missions, including:

- DCA   Defensive Counterair, defense against penetrators
- Sweep   Fighters that precede penetrators to engage DCA
- SEAD   Suppression of Enemy Air Defenses, attacking SAMs
- OCA   Offensive Counterair, attacking air bases
- AI   Air Interdiction, attacking fixed targets.

Some of the factors accounted for in the model are:

- Sortie rates by aircraft type and basing distance, including surge capability
- Air-to-air missile types and numbers carried
- Precision-guided munition numbers and capabilities
- Aircraft capabilities and stealth differences
- Pilot quality

- Differing mission packages
- Geographic and mission effects on engagement rate
- Engagements per sortie (and the advantages of fire-and-forget missiles)
- First-shot advantages for more modern aircraft
- Suppressed or aborted DCA and strike sorties beyond attrition
- Ballistic missile attacks on SAMs and air bases
- Effect of IADS damage on SAMs
- Effect of air base damage on sortie generation
- Aircraft killed in shelters
- Air base repair.

Air war adjudication is done in the following steps:

- Sweep packages attacking DCA
- SEAD packages attacking SAMs
- SAMs attacking sweep and strike packages
- DCA packages attacking strike package escort
- DCA packages attacking strike package mission sorties.

Because the JICM is a deterministic model, engagements are not resolved by randomly pairing packages but by fighting each attacking package against each defending package. Probabilistic factors that would normally select some packages but not others (such as the percentage of packages engaging) are instead applied as fractions to each package.

The same general process applies to each of the adjudication steps above. First, engagement rates determine the number of sorties from each package that engages. Second, each attacking package is allocated across all defending packages such that defending packages each face a composite attacker made up of slices of each attacking package. Third, each defending package is adjudicated against its composite attacker. The engagement vulnerability score of each package determines the fraction of each side that shoots before getting shot at. Attrition is then totaled and allocated back to the participating packages.

AIR-TO-AIR ADJUDICATION

Engagement Rates

Engagement rates limit the number of sorties that engage in each adjudication step and are calculated for each package. Because the JICM is a deterministic model, engagement rates are applied separately to each package. For example, an engagement rate of 75 percent means that 75 percent of the sorties in a package engage rather than 75 percent of the packages.

The engagement rate for attacking packages is based on factors that represent how well the attacker can predict and cover the areas where the defender will be and how well the attacker can react when he is less than perfect in the first factor.

The Taiwan theater is small and as a relatively few air bases are the focus of the strikes, the coverage factor for sweep attacking DCA is a fixed 100 percent. Sweep does not have a reaction adjustment because it must largely predict where DCA will appear.

For DCA attacking strike packages, the coverage factor is based on the physical space the total DCA sorties can cover compared with the size of the theater. Again, because of the small size of the theater, this factor was always 100 percent. The reaction adjustment for DCA is represented by raising the coverage fraction to the exponent 0.5.

Engagement rates for defending packages are based on the coverage of the attacking packages and the vulnerability of the defending aircraft to engagement. Vulnerability is a data item for each aircraft type that represents the ability of the aircraft to avoid engagement through a combination of stealth, avionics, performance, and weapon range. Typical values are from 0.5 for modern aircraft to 1.0 for prior-generation aircraft.

For sweep attacking DCA:

$$\text{base-sweep-engagement-rate} = \text{sweep-coverage-parameter}$$

$$\text{base-DCA-engagement-rate} = \text{sweep-coverage-parameter} \bullet \text{package-vulnerability}$$

For DCA attacking strike packages:

$$\text{base-DCA-engagement-rate} = \text{DCA-theater-coverage} \wedge 0.5$$

$$\text{base-striker-engagement-rate} = \text{DCA-theater-coverage} \bullet \text{package-vulnerability}$$

Note that the engagement rates of all attacking packages will be the same, while those of the defending packages will vary according to their individual vulnerabilities.

An additional engagement constraint is based on the ratio of total engaged sorties. Because DCA is trying to avoid sweep and sweep must spread out to cover DCA operating areas, we limited the ratio of total engaged sweep and DCA sorties to 1:1, meaning that each sweep sortie engaged at most one DCA sortie. We limited the ratios of DCA versus escort and strike sorties to 2:1, meaning that up to two DCA sorties could engage each escort or strike sortie.

$$\text{sortie-ratio} = \frac{\text{total-friendly-engaged-sorties}}{\text{total-enemy-engaged-sorties}}$$

For each package:

$$\text{engagement-rate} = \text{base-engagement-rate} \bullet \text{minimum}(\text{sortie-ratio}, \text{allowed-ratio})$$

Attacker Allocation Across Defenders

The attacker allocation process allocates a fraction of each attacking package to each defending package. For air-to-air engagements, the allocation is in proportion to the number of engaged sorties in the defending package compared with the total for the side.[1]

For each attacking package versus each defending package:

$$\text{fraction-allocated} = \text{defender-sorties} \bullet \frac{\text{package-engagement-rate}}{\text{total-engaged-sorties}}$$

For example, consider an engagement with two sweep packages of four sorties, each with an engagement rate of 50 percent, versus two DCA packages of four sorties with engagement rates of 100 percent and 50 percent. Two sorties in each sweep package are engaged, versus four sorties in the first DCA package and two sorties in the other. Therefore, four-sixths of each sweep package is allocated against the first DCA package, and two-sixths against the second.

Adjudicating Air-to-Air Engagements

Following the allocation of attackers to defenders, each defending package is adjudicated in turn against the collection of fractional packages allocated against it.

The adjudication process depends on the number and score of weapons carried and the air-to-air vulnerability of each sortie. The defending package (DCA, striker escort, or striker mission sorties) is always a uniform type of aircraft, and so their true weapons and vulnerability are used. The weapons used by the attacker are the total across all allocated sorties but with an averaged score. The vulnerability of the attacker sorties is taken to be the best vulnerability of any of the allocated sorties.

In our cases, PRC sweep, striker escort, and striker mission sorties were of uniform vulnerability, but ROC DCA sorties were not. The ROC IDF aircraft has a worse vulnerability than other ROC aircraft (0.8 compared with 0.5). This results in a bias in favor of the ROCAF

---

[1] For engagements involving disparate types of forces, such as SAMs engaging aircraft, missiles, and standoff weapons, the effectiveness of the attacker against the different defenders and the ability of the attacker to discriminate between defenders is also considered.

68   Dire Strait?

when ROC DCA is attacking strike packages, because all sorties use the best (0.5) vulnerability. Because the PRC striker packages are previous-generation aircraft, this engagement is one-sided anyway.

For each defending package:

attacker-shots = total from allocated attacking sorties

attacker-score = average from attacker weapons

attacker-vulnerability = best from allocated attacking sorties

defender-shots = total from defender sorties

defender-score = score from defender weapons

defender-vulnerability = vulnerability from defender sorties.

Shots Taken

Because of the short sortie distances in this theater and the training limitations of the two sides, we limited each sortie carrying semiactive guidance missiles to only one engagement. To accomplish this, we loaded these aircraft with only two missiles, which would both be fired at the first target engaged. BVR fire-and-forget weapons, such as the AIM-120 or AA-12, were not subject to this constraint, and sorties were allowed their normal complement of these weapons in addition to two semiactive guidance missiles. Such aircraft as the U.S. F-15C, which carried four AIM-120s and two AIM-9, could shoot at up to three targets with two missiles each.

maximum-shots-per-shooter = 2 + fire-and-forget-weapons.

The number of shots taken at each target was also limited to four, but reduced by the target sortie's air-to-air vulnerability.

maximum-shots-per-target = 4 • air-to-air-vulnerability.

Therefore, modern aircraft with 0.5 vulnerability could have at most two shots taken against them, while previous-generation aircraft could have up to four.

This constraint was only limiting in the sweep versus DCA engagement when sorties with fire-and-forget missiles engaged modern,

0.5-vulnerability aircraft. In these cases, the shots by the ROC DCA that could not be expended against the PRC sweep were later used against the 1.0-vulnerability escorts and strikers. Because the ROC DCA also includes 0.8-vulnerability IDF aircraft, PRC sorties with AA-12s are generally not shot-limited.

Weapon scores represent an expected number of kills (EK). The score for each shot is an average of all weapons on the sortie.

First Shot

In addition to limiting the number of engaged sorties, air-to-air vulnerability also determines the fraction of each side that is given the first shot. The adjudication of each defending package against its allocated attackers begins with determining the fraction of first shots that goes to the side with the lowest vulnerability.

first-shot-fraction = 0.5 + 0.5 • (high-vuln – low-vuln) ^ 0.2.

Figure B.1 shows the relationship between the difference in vulnerability and the percentage of first shot that goes to the side with the

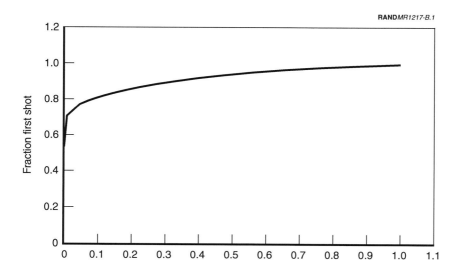

Figure B.1—Fraction of Package with First Shot

lower vulnerability. Note that at equal vulnerability (a difference of zero) 50 percent of each side shoots first.

This fraction is implemented by dividing the adjudication into two subadjudications according to the first-shot fraction. For example, a vulnerability difference of 0.1 gives a first-shot fraction of 0.82. In the first adjudication, 82 percent of the side with the lowest vulnerability shoots first at 82 percent of the other side, and the survivors shoot back. Then 18 percent of the side with the higher vulnerability shoots first at the other side, and the survivors shoot back. The final attrition is the sum of the attrition from both steps.

In the following example, side A is the side with lower vulnerability.

First adjudication step:

$$\text{attrition}_B = \text{first-shot-fraction} \bullet \text{engaged-sorties}_A$$
$$\bullet \text{shots-per-sortie}_A \bullet \text{EK-per-shot}_A$$

$$\text{attrition}_A = \text{first-shot-fraction} \bullet (\text{engaged-sorties}_B - \text{attrition}_B)$$
$$\bullet \text{shots-per-sortie}_B \bullet \text{EK-per-shot}_B$$

Second adjudication step:

$$\text{attrition}_A = (1-\text{first-shot-fraction}) \bullet \text{engaged-sorties}_B$$
$$\bullet \text{shots-per-sortie} \bullet \text{EK-per-shot}_B$$

$$\text{attrition}_B = (1-\text{first-shot-fraction}) \bullet (\text{engaged-sorties}_A - \text{attrition}_A)$$
$$\bullet \text{shots-per-sortie}_A \bullet \text{EK-per-shot}_A$$

Allocating Attrition

Because the adjudication process fights each entire defending package against a fraction of each attacking sortie, defender attrition is simply posted against that package. Attacker attrition is allocated back to all attacking packages in proportion to their allocation divided by the average vulnerability of all attacking sorties. Thus, sorties that are more or less vulnerable than the average will take more or less attrition.

$$\text{attacker-package-attrition} = \text{total-attrition} \bullet \text{allocation-fraction}$$
$$\bullet \frac{\text{package-vulnerability}}{\text{average-vulnerability}}$$

Because this is a deterministic model, both fractional shots and fractional kills are allowed.

Sortie Suppression

For every DCA sortie killed by sweep or escort, an equal number of sorties are suppressed or rendered incapable of engaging strike packages. This represents DCA that has been drawn out of position by the sweep without getting shots.

BIASES IN THE JICM REPRESENTATION OF THE AIR WAR

Limiting sorties without AIM-120s or AA-12s to one engagement discounts the value of modern aircraft that carry as many as eight missiles. We felt that engagement opportunities in a small theater would be limited, with flight times from the edge of the PRC SA-10 envelope to targets no more than 20 minutes for the slowest aircraft. Conversely, the impact of BVR fire-and-forget missiles is magnified, with sorties carrying these weapons able to get as many as three times the kills as sorties without them.

The restriction that each sweep sortie engage no more than one DCA sortie is another potential bias. Model and time limitations prevented us from looking at a wider range of engaged sortie ratios, as well as uncertainty in how shot opportunities change in these cases. This ratio is as much a function of the DCA's ability to avoid the sweep as anything else, and could fall below 1:1 as well as rise above it. Lower ratios would allow more DCA to get to strikers, while higher ratios would result in more sweep attrition.

When ROC DCA engaged strike package escorts, we allowed up to a 2:1 engaged sortie ratio to represent the fact that DCA cannot evade escorts as easily. This allowed up to four shots to be taken at each escort, while allowing only two in return, resulting in extremely high escort attrition. While possibly biased against the escorts, we felt the disparity between DCA and escort quality made this a reasonable outcome.

The engaged sortie ratio limits also reduce the impact of changes in sortie quantities. Attrition in the sweep versus DCA engagement is essentially linear with the number of sorties on the smaller side—normally the sweep but in some cases the DCA. While the unengaged DCA goes on to engage the strike packages, the PRC gets no benefit from unengaged sweep. In cases, however, in which the PRC has unengaged sweep, it is already winning the air war.

Aircraft quality enters into the model in a number of ways. As a training factor, quality multiplies lethality and has a linear effect on the sweep versus DCA engagement and up to a squared effect on DCA versus escorts. BVR missiles raise lethality by a factor of three but have more of an impact on ROC DCA that has more opportunities for engagements. DCA sorties that are limited to firing two shots at 0.5-vulnerability sweep sorties can fire the additional shots at escort and strikers. PRC AA-12 shooters have only one opportunity each to engage DCA and are limited to two shots against 0.5-vulnerability F-16As and Mirage 2000s, but can take up to 3.2 shots against 0.8-vulnerability IDFs. Still, in cases with advanced missiles the ROC DCA sorties will usually get more shots than the PRC sweep.

Aircraft air-to-air vulnerability is another measure of quality. It linearly reduces the number of sorties engaged and the shots that can be taken at a sortie but has a nonlinear effect on the fraction of sorties that get first shot, killing before being killed. We categorized the aircraft in the scenario into three vulnerability groups spaced well apart. Because of the shape of the first shot equation, small changes in vulnerabilities have little effect on aircraft that are already sizably different. Therefore, scenario outcomes are not strongly driven by small changes in the vulnerability scores of these groupings.

AIR DEFENSE ADJUDICATION

SAM engagements with air packages are adjudicated by a process similar to that for air-to-air engagements, with each SAM battery treated as an attacking package. Engagement rates are determined for both sides, attacking SAMs are allocated across defending packages, first shot calculations are made, and attrition assessed.

IADS Model

Since the JICM does not have an explicit model for the support of SAM batteries by an integrated air defense system (IADS), we implemented a simple parameter model in the JICM order scripting language. This model reduced ROC SAM effectiveness by half the percentage of damage done to Taiwan's 10 modeled early warning radar sites. In all our cases, these radars were targeted by 20 DF-21 missiles and destroyed before the first air strike, resulting in a 50 percent effectiveness penalty to the ROC SAMs.

SEAD

In all our cases, PRC air did not fly SEAD missions because its most capable aircraft were involved in the air-to-air battle. Instead, TBMs with cluster warheads were fired at SAMs. In this analysis, we assumed that the ROC Patriots could not effectively intercept the modern PRC missiles. We also assumed that only 50 percent of the ROC SAMs were targetable by missiles on any given day due to movement or decoy measures.

OCA AND AI ADJUDICATION

On-Target Air-to-Ground Sorties

For each sortie lost on ingress another sortie aborts before attacking the target. This represents the loss in effectiveness caused by flying in an intense threat environment.

$$\text{abort-rate} = \text{loss-rate}$$

Air Base Attack

We modeled each of Taiwan's six air bases that support tactical aircraft as having one runway and four maintenance sites. In the JICM, air base sortie generating capability degrades as a function of damage to runways and maintenance sites, with a minimum (20 percent) below which the capability cannot be reduced.

$$\text{sortie-generation-rate} = 0.4 \bullet \text{maintenance-survival} + 0.2$$

This formulation requires attacks on both runways and maintenance to severely limit operations.

Air base repair is calculated according to the exponential functions:

$$\text{runway-percent-surviving} = 0.98 \bullet (1.0 - e \wedge (-0.1 \bullet t))$$

$$\text{maintenance-percent-surviving} = 0.90 \bullet (1.0 - e \wedge (-0.01 \bullet t)),$$

where t is the time spent repairing.

Maintenance repair is nearly linear, repairing at approximately 10 percent per day, to a maximum of 90 percent. Runway repair is more strongly nonlinear, repairing more than 30 percent in the first day when completely cut.

SCENARIO

ATO Creation

The JICM creates an ATO at the start of each day by assembling sorties into air-to-air and strike packages according to provided package definitions and other planning guidance. Table B.1 lists the package definitions for the missions used in this scenario.

Table B.1

Mission Packages

Side	Mission	Packages
U.S.	DCA	4 F-15C or F/A-18E/F
ROC	DCA	4 F-16A/Mirage 2000/IDF
PRC	DCA	4 J-7
	Sweep	4 Su-27/Su-30/J-10
	OCA or AI	4 H-6 and 4 escort
		8 JH-7/Q-5/J-7

Escorts for PRC strike packages were taken first from previous-generation fighters, leaving the modern PRC fighters for the sweep mission. J-7s flexed as required to fill out strike or escort roles.

Packages created at the start of the day were scheduled across the six four-hour periods. Table B.2 lists the percentages of packages by period.

Twenty-five percent of U.S. F-15C sorties and U.S. carrier sorties were withheld to provide air base and carrier defense. The small U.S. presence during night periods (1, 2, 6) was assumed to be flying combat air patrol (CAP) and escorting reconnaissance aircraft (not explicitly modeled). The U.S. F-15Cs out of Okinawa maintain a level effort CAP during the day periods because they are based too far away to be completely reactive to PRC strikes. U.S. carrier air, which is closer, can concentrate more sorties in the two PRC strike periods.

ROC and U.S. land-based F-15Cs were allowed to surge to 150 percent of their base sortie rates for the first 48 hours of combat. PRC sorties surged 125 percent also for 48 hours, while U.S. carrier-based sorties did not surge.

Air-to-Air Combat

Each day's air combat was fought in three periods, with strikes by the PRC in periods 3 and 5 and a smaller fighter sweep in between. With the base threat, the PRC strikes consist of approximately 90 sweep sorties, followed by 250 OCA and AI sorties with 90 escorts. With the advanced threat, the added advanced fighters boost sweep to 200 sorties per strike, while reducing OCA and AI to 200 sorties. ROC

Table B.2

Package Timing

Side	Mission	Period 1	2	3	4	5	6	Withheld
U.S.	DCA (F-15)	5	5	20	20	20	5	25
	DCA (F/A-18)	5	5	27	5	28	5	25
ROC	DCA	0	0	40	20	40	0	0
PRC	Sweep	0	0	40	20	40	0	0
	Strike	0	0	50	0	50	0	0

DCA flew 400 sorties against the first strike but was reduced by air base damage and attrition to 250 against the second.

Figure B.2 shows the D-day first strike sorties for the base case.

Figure B.3 shows the resulting sortie losses for each of the three periods, from air-to-air and ground-to-air (SAMs) combat.

Figure B.4 shows total aircraft losses and total sorties flown on D-day by aircraft type. The percentage of sortie losses is shown above the bars.

Ballistic Missiles

In the base case, we assumed that the PRC would launch the bulk of its missiles in two strikes on the first day of combat. Each missile strike preceded an air strike, for maximum effect on defending sorties. Twenty DF-15 missiles with both cluster and GPS-guided high-explosive warheads were fired at each of the six air bases with tactical

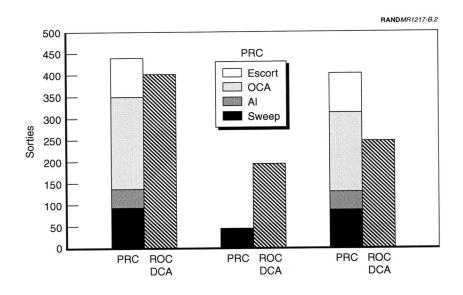

Figure B.2—D-Day Sorties

Overview of the JICM 77

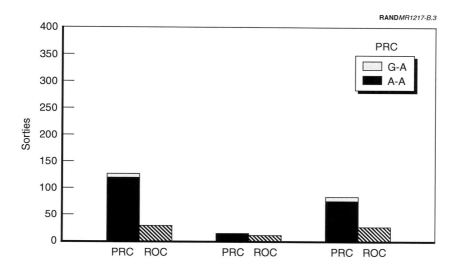

Figure B.3—D-Day Sortie Losses

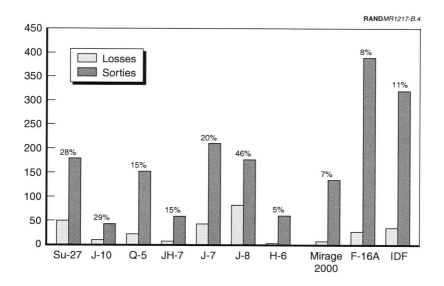

Figure B.4—D-Day Aircraft Losses and Sorties

aircraft, dropping air base sortie generation to 20 percent by the end of D-day (although overnight repairs raised sortie generation to 33 percent). Remaining DF-15s with cluster warheads were fired at known Patriot and Tien Kung SAM sites, killing six batteries on D-day. Twenty DF-21 missiles were fired at early warning radars, killing 10 sites and dropping ROC SAM effectiveness by 50 percent. DF-11 missiles were fired at landing preparation sites that were not explicitly modeled.

In cases with increased numbers of missiles, additional DF-15s with GPS guidance were used to restrike air base runways for two additional days, while additional DF-15s were used against SAM sites, killing more than 30 SAM batteries in four days.

Air-to-Ground Strikes

During the scenario, 80 percent of air strikes were directed against air bases. Dumb bombs were dropped against air base maintenance sites, GPS-guided munitions against runways, and laser-guided munitions with penetrating warheads against aircraft shelters. In the base case, high-altitude bombing with dumb weapons was largely ineffective and the numbers of PGMs were insufficient to change the outcome. In cases with increased numbers of PGMs, GPS-guided munitions were used both on runways and maintenance sites and were generally capable of reversing repair efforts. Because the missile attacks grounded many sorties, LGB attacks on shelters were reasonably effective, killing in the increased-munition, advanced-threat cases up to 35 aircraft in four days. In many cases, however, the ROCAF could prevent most of the strikes from reaching their targets.

Data

Table B.3 shows the key data used by the air model for Taiwan, PRC, and U.S. aircraft.

Sortie rate is a single number for each type of aircraft. It is not varied by the mission flown, although sortie rate multipliers are set for the region in which the squadron is based. In this case, the sortie rate for the U.S. F-15C is reduced because these aircraft are based in Okinawa.

Table B.3
Aircraft Data

			Vulnerability	
Type	Sortie Rate	Air-to-Air Multiplier	Air-Air	Ground-Air
Mirage 2000	2.0	.80	0.5	—
F-16A	2.0	.85	0.5/0.4	—
IDF	2.0	.85	0.8	—
Su-27	2.0	.90	0.5/0.4	0.5
Su-30	2.0	.90	0.5/0.4	0.5
J-10	1.5	.80	0.5/0.4	0.5
Q-5	1.0	—	1.0	1.0
JH-7	1.0	—	1.0	0.5
J-7	1.0	.80	1.0	1.0
J-8	1.0	.80	1.0	1.0
H-6	1.0	—	1.0	1.0
F/A-18E/F	2.0	.85	0.4	0.5
F-15C	1.6	.80	0.4	0.5

The **air-to-air multiplier** represents how the airframe of each aircraft type modifies the lethality of the weapons it carries.

Vulnerability represents how difficult the aircraft is to engage, both in air-to-air and ground-to-air combat. It primarily represents stealthiness, but it also includes performance, avionics, and weapon range. Vulnerabilities following slashes are the vulnerabilities employed when the aircraft carries an AIM-120 or AA-12 missile. Vulnerability has three effects in the model: it reduces the number of sorties that are engaged; it reduces the number of shots taken at the sortie; and it determines what fraction of sorties get first shot in an engagement.

We divided aircraft into four groups: modern aircraft with AIM-120s or AA-12s, other modern aircraft, previous-generation aircraft, and the Taiwan IDF. Table B.4 shows the engagement rates and first shot percentages derived for these groups.

Table B.4

Engagement Rates and First Shots

Type	Air-to-Air Vulnerability	First Shot Percentage Versus Adversary of Vulnerability			
		0.4	0.5	0.8	1.0
Modern with BVR	0.4	50.0	82.0	92.0	95.0
Modern	0.5	18.0	50.0	89.0	94.0
IDF	0.8	8.0	11.0	50.0	86.0
Old	1.0	5.0	6.0	14.0	50.0

Table B.5 shows the lethality scores for the air-to-air weapons used. The number of air-to-air weapons was unconstrained, except for Taiwan's 240 MICA missiles, which were exhausted after two days of combat. AIM-120 and AA-12 missiles were available to Taiwan and the PRC in some cases.

Air-to-air weapon lethality is represented as single-shot EK. These are not test range numbers, but rather the scores represent an average lethality across the kinds of engagements occurring in an air campaign. We chose to divide these air-to-air weapons into three categories for scoring: BVR missiles, other modern missiles, and the AA-2.

Table B.5

Air-to-Air Weapon Data

Type	EK
AA-12	.70
AIM-120	.70
MICA	.70
AA-10	.35
AA-11	.35
AIM-9	.35
AIM-7	.35
AA-2	.17

Training factors are represented as multipliers on weapon EK, shown in Table B.6.

Weapon loads for each air-to-air mission are shown in Table B.7. Where more than one load is shown for a mission, loads are listed in order of preference.

The model uses an average EK per shot across the entire weapon load. The average EKs shown above also include the training factor and air-to-air multiplier.

Table B.6

Training Factors

Side	Factor
U.S.	1.0
Taiwan	0.8
PRC	0.5

Table B.7

Air-to-Air Weapon Loads

Type	Air-to-Air Loads	Shots	EK per Shot
Mirage 2000	2 MICA	2	0.45
F-16A	2 AIM-7	2	0.24
	2 AIM-9, 4 AIM-120	6	0.39
IDF	2 AIM-7	2	0.24
J-7	2 AA-2	2	0.06
J-8	2 AA-10	2	0.14
J-10	2 AA-10	2	0.14
	2 AA-11, 2 AA-12	4	0.21
Su-27	2 AA-10	2	0.15
	2 AA-11, 4 AA-12	6	0.26
Su-30	2 AA-10	2	0.15
	2 AA-11, 4 AA-12	6	0.26
F/A-18E/F	2 AIM-9, 4 AIM-120	6	0.49
F-15C	2 AIM-9, 4 AIM-120	6	0.46

Given the first shot percentages, exchange rates for selected aircraft are shown in Table B.8 as a point of comparison with mission-level models.[2]

Table B.9 shows the number and lethality scores for the air-to-ground weapons used by the PRC. Where there are two numbers for quantity, the larger number was used in cases with increased availability of PGMs or ballistic missiles. In the PGM-limited cases, GPS and LGBs are used up in two days of strikes against air bases. Only half the LGBs had penetrator warheads capable of busting shelters.

Air-to-ground lethality is represented as EKs against standard types of targets. Hard targets in this scenario are aircraft shelters, soft targets are early warning radars and landing preparation targets, area targets are air base maintenance sites, runways are air base runways, and SEAD targets are SAM batteries. SAM kills represent the kill of a single critical element, such as the radar or control vehicle. We assumed that there would be no reconstitution of SAM batteries within a four-day combat.

All air-to-ground attacks were made from high altitude to avoid short-range air defense systems.

Weapon loads for air-to-ground missions are shown in Table B.10. Where more than one load is shown for a mission, loads are listed in order of preference. Total load EKs are given for OCA against shelters, runways, and maintenance facilities; for AI against radars and

Table B.8

Air-to-Air Exchange Rates

Type	Su-27 with AA-12	Su-27	J-8
U.S. F-15C with AIM-120	2.1	6.3	53.5
ROC Mirage 2000 with MICA	1.2	3.6	5.5
ROC IDF	0.3	1.4	2.1

---

[2]These values were calculated by going through the attrition process with a single four-sortie package on each side.

Table B.9

Air-to-Ground Weapon Data

Weapon	Quantity	EK Hard	EK Soft	EK Area	EK Runway	EK SAM
GPS-guided (800-kg)	200/2,000		0.71	0.12	0.06	
Laser-guided (800-kg)	50/500	0.35	0.65	0.25		
Cluster (500-kg)	—		0.01	0.007		
Dumb (250-kg)	—	0.03	0.005	0.004		
DF-21	80/160		1.00			
DF-11	50/100		1.00			
DF-15 cluster	80/200			0.50		0.33
DF-15 GPS-guided	80/120				0.08	

landing preparation sites; and for SEAD against SAM batteries. These EKs include the training factor from Table B.6.

Table B.11 shows the number of ROC SAM batteries and their EKs versus high-altitude aircraft. In all cases, we assumed that the ROC Patriots, which were sited at air bases, could not intercept the more modern missiles that were fired at them.

Table B.10

Air-to-Ground Weapon Loads

Aircraft	Load for OCA	EK Shelter	EK Runway	EK Maintenance
Q-5/J-7	4 dumb			0.016
JH-7	2 GPS		0.12	0.24
	2 LGB	0.76		0.50
	4 dumb, 4 cluster			0.044
H-6	12 dumb			0.048
	3 GPS		0.36	

Table B.11

Ground-to-Air Weapon Data

Type	Quantity (Battery)	EK versus Aircraft
Patriot PAC-2	9	0.7
Tien Kung	6	0.7
Hawk	36	0.4

REFERENCES

Allen, Kenneth W., Glenn Krumel, and Jonathan D. Pollack, *China's Air Force Enters the 21st Century*, Santa Monica, Calif.: RAND, MR-580-AF, 1995.

Anderson, James H., *Tensions Across the Strait: China's Military Options Against Taiwan Short of War*, Heritage, Washington, Backgrounder No. 1328, September 1999.

Bitzinger, Richard A., Bates Gill, *Gearing Up for High-Tech Warfare?* Washington, D.C.: Center for Strategic and Budgetary Assessments, 1996.

Cheung, T. M., "Chinese Military Preparations Against Taiwan Over the Next 10 Years," in James R. Lilley and Chuck Downs, eds., *Crisis in the Taiwan Strait*, Washington, D.C.: National Defense University Press, 1997.

Cullen, C., and C. F. Foss, eds., *Jane's Land-Based Air Defence 1997–98*, Coulsdon, U.K.: Jane's Information Group, 1997.

Daly, M., "Democracy Is Taiwan's Best Shield Against China's Threat," *Jane's International Defense Review*, Vol. No. 32, April 1999.

Dreyer, J. T., "China's Military Strategy Toward Taiwan," *The American Asian Review*, Fall 1999.

Fisher, R. D., Jr., "China's Missiles Over the Taiwan Strait: A Political and Military Assessment," in James R. Lilley and Chuck Downs, eds., *Crisis in the Taiwan Strait*, Washington, D.C.: National Defense University Press, 1997.

Fulghum, D. A., "Israel Builds China's First AWACS Aircraft," *Aviation Week & Space Technology*, Vol. 151, No. 22, November 29, 1999.

Gertz, B., "China Moves Missiles in Direction of Taiwan," *Washington Times*, February 11, 1999, p. A12.

International Institute for Strategic Studies (IISS), *The Military Balance 1998/99*, London: Oxford, 1998.

Jackson, P., ed., *Jane's All the World's Aircraft 1998–99*, Coulsdon, U.K.: Jane's Information Group, 1998.

Jane's Information Group, *Jane's World Air Forces*, Coulsdon, U.K., 1998.

Jones, Carl M., and Daniel B. Fox, *JICM 3.5: Documentation and Tutorials*, Santa Monica, Calif.: RAND, DRR-2081-OSD, April 1999.

Lennox, D., ed., *Jane's Strategic Weapons Systems*, Coulsdon, U.K.: Jane's Information Group, 1999.

Lin, Yuan, "Sea-Crossing Offensive Capability of Chinese Armed Forces," in *Hong Kong Kuang Chiao Ching*, August 16, 1997, translated in Foreign Broadcast Information Service FBIS-CHI-97-268.

McFarland, Stephen Lee, Wesley Phillips Newton, and Richard P. Hallion, *To Command the Sky: The Battle for Air Superiority Over Germany, 1942–1944*, Washington, D.C.: Smithsonian Institution Press, November 1991.

Nordeen, Lon, *Fighters Over Israel*, New York: Orion Books, 1990.

Novichkov, N., "Four Sovremennys in Total for Beijing," *Jane's Defence Weekly*, Vol. 33, No. 11, March 15, 2000.

Sae-Liu, R., "Su-30MK Purchase on Chinese Agenda," *Jane's Defence Weekly*, Vol. 32, No. 6, August 11, 1999.

Schelling, T. C., *Arms and Influence*, New Haven, Conn.: Yale University Press, 1966.

Sharpe, Richard, ed., *Jane's Fighting Ships 1998–99*, Coulsdon, UK: Jane's Information Group, 1998.

Stillion, John, and David Orletsky, *Airbase Vulnerability to Conventional Cruise-Missile and Ballistic-Missile Attacks: Technology, Scenarios, and U.S. Air Force Responses*, Santa Monica, Calif.: RAND, MR-1028-AF, 1999.

Taylor, M. J. H., ed., *Jane's World Combat Aircraft*, London: Jane's Information Group, 1988.

Terraine, John, *A Time for Courage: The Royal Air Force in the European War, 1939–1945*, New York: Macmillan, 1985.

U.S. Central Intelligence Agency, "Taiwan," *World Factbook 1998*, http://www.odci.gov/cia/publications/factbook/tw.html, June 9, 1999.

U.S. Congress, 96th Congress, First Session, *Taiwan Relations Act*, Public Law 96-8, section 3 (a), 1979.

U.S. Department of Defense, *Selected Military Capabilities of the People's Republic of China*, Washington, D.C., 1997.

U.S. Naval Institute, *Periscope* on-line USNI Military Database, URL http://www.periscope.ucg.com, November 1999.

U.S. Secretary of Defense, *The Security Situation in the Taiwan Strait*, Report to Congress, Washington, D.C., February 1999.

Wang, P., *Military Matchups: PRC vs. ROC*, http://www.emeraldesigns.com/matchup/military.shtml, November 1999.

World Navies Today, http://www.uss-salem.org/worldnav/, 1998–1999.